Peter Spear Dingey

Machinery pattern making containing full size profiles of gear teeth : and fine engravings on full-page plates, illustrating manner of constructing numerous and important patterns and core boxes

Peter Spear Dingey

Machinery pattern making containing full size profiles of gear teeth : and fine engravings on full-page plates, illustrating manner of constructing numerous and important patterns and core boxes

ISBN/EAN: 9783337156749

Printed in Europe, USA, Canada, Australia, Japan

Cover: Foto ©berggeist007 / pixelio.de

More available books at **www.hansebooks.com**

MACHINERY PATTERN MAKING

CONTAINING

FULL SIZE PROFILES OF GEAR TEETH

AND

FINE ENGRAVINGS ON FULL-PAGE PLATES, ILLUSTRATING MANNER OF
CONSTRUCTING NUMEROUS AND IMPORTANT
PATTERNS AND CORE BOXES

BY

P. S. DINGEY

PRACTICAL PATTERN MAKER AND MECHANICAL DRAFTSMAN

376 Fine Illustrations

NEW YORK
JOHN WILEY & SONS
53 E. TENTH STREET

COPYRIGHT, 1891
BY P. S. DINGEY

The Caxton Press
171, 173 Macdougal Street, New York

NOTE.

Most of the matter in this book was written expressly for the *American Machinist*, to whose courtesy we are indebted for some of the illustrations. Mr. Dingey has, however, revised many of his drawings and much of the matter, adding some valuable items.

WILEY & SONS.

PREFACE.

It is assumed that those who will read this book do not need the rudiments of pattern making presented, therefore the elementary part of the business, which to most pattern makers is so distasteful, has been omitted. The author has not laid down any cast iron rules as to the methods set forth of doing work, and desires that the contents be accepted as suggestions; at the same time it must be understood that he has not shown a number of ways and means for experimental purposes, but that which is given is practicable, and the result of practice, and of over twenty years experience in the business.

The object of this book has not been to teach pattern making, for that can never be done through a book, but to discuss methods.

There are some who regard with a great deal of jealousy anything that comes to them about the practical part of their trade through a book. Then there are others that scarcely ever open a book to read or study, that condemn all books, treating them with contempt, and passing sentence on the writers who dare write on anything, the fringe of which their small minds might have grasped. For the latter this book is

PREFACE.

not intended. With the former there is some reason for the feeling that exists in their minds, and it may be found in the fact, that, from time to time, there have been good writers, men of intelligence, those whom we should always honor and respect, who have written on the practical, when only possessing the theoretical knowledge of their subject.

In the present volume everything of a visionary kind has been avoided, and the author has presented such subjects as he believes will be interesting to pattern makers and those of the machinery business generally.

P. S. DINGEY.

CHICAGO, ILL.

CONTENTS.

THE PATTERN MAKER AND HIS TRADE,	1
THE PATTERN SHOP—Its Position, Size, and Requirements,	5
MARKING AND RECORDING PATTERNS,	9
PRINTING-PRESS CYLINDERS,	13
DIFFERENTIAL CHAIN PULLEYS,	15
A HANDY TOOL FOR LAYING OUT HEXAGON NUTS,	18
HOW TO CAST JOURNAL BOXES ON FRAMES,	19
HOW TO STRIKE AN ARC BY THE AID OF THREE POINTS,	20
KEY-HEADS FOR MOTION RODS—The way to lessen the cost of their production,	22
ELBOW AND TEE PIPES—A quick method for turning the patterns aud core-boxes in the Lathe,	24
SLIDE VALVE CYLINDERS,	26
CORLISS CYLINDERS—With a full description, showing how to construct patterns and core-boxes which can be changed at short notice for different stroke Engines,	29
FLY WHEELS—Different styles,	36
ENGINE FRAMES—How to build the pattern to serve for various strokes,	40
SPUR GEARS—How the teeth should be made,	44
BEVEL GEARS—The manner of laying them out,	48
HOW TO LAY OUT THE THREAD OF A WORM FOR THE PATTERN,	51
WORM WHEELS—The way to get the angle of teeth and the manner of fastening them on,	53

CONTENTS.

SWEEPING STRAIGHT WINDING DRUMS,	56
MAKING WINDING DRUMS FROM PATTERNS—Method of cutting the groove,	58
MAKING SHEAVES FROM CORE-BOXES,	60
MAKING SHEAVES FROM PATTERNS,	65
SHEAVES WITH WROUGHT IRON ARMS—An original way of making the Hub,	68
A MACHINE FOR SWEEPING CONICAL DRUMS—Designed by the author,	70
GEAR TEETH—One hundred and Twenty-eight full size different profiles of Gear Teeth from 1″ to 3″ Pitch, suitable for gears having from 14 to 800 teeth,	74
Table showing at a glance the required diameters of Gear Wheels for a given number of teeth and pitch,	76, 77, 78, 79, 80, 81, 82
Weight of Cast Iron Pipe,	83, 84
" " Cast Iron Balls,	83
" " Round Cast Iron,	85
" " Square " "	85
" " Flat " "	86
" " Superficial Foot of Cast Iron from ¼″ to 2″ thick	86
" " Round Lead,	87
" " Square "	87
Binary and Decimal Fractions,	87
Table which gives distances to open a 2 ft. rule for obtaining angles from 1° to 90°,	88
Metric Measure reduced to inches,	89

THE PATTERN MAKER AND HIS TRADE.

A THEORETICAL knowledge of moulding, with an ability to read drawings well, are indispensable to a good pattern maker. He has to know how the pattern is to be moulded before he can do much, and to see the machine, or parts of it, mentally, just as the draftsman sees it.

In many trades, that which is most necessary is to become an expert in handling the tools. This is not so in pattern making. There is something far more important than merely cutting wood.

In many patterns it is not so much a question of workmanship, as knowledge. A pattern, after it is made, may be duplicated by any ordinary wood worker; fine workmanship may not have been the all-important, and yet none but a first-class pattern maker could have planned and made it.

On the other hand, there is much that calls for fine workmanship and less scheming. This is no doubt true, more or less, in all trades, but it is especially so in pattern making, and this is why I say that pattern making is not merely cutting wood. From the very nature of the trade, a pattern maker is a good worker in wood, because he is accustomed to work to finer measurements than the ordinary wood worker.

I think the responsibility that rests upon the pattern

department, as to whether work turns out right, is equal to that of the drawing room; for while the draftsman is responsible for the design, upon the pattern maker rests a large proportion of the responsibility of executing correctly that which has been put upon paper.

The liability to mistakes is reduced considerably when the machinist takes hold where the pattern maker has left off; the machinist's part is no doubt the most important as to the workmanship and right working of the machinery; he can make it good, bad, or indifferent; but mistakes in measurements he is not so liable to as the pattern maker, because the machinist has the casting, and is given the drawing of it with instructions to finish to drawing.

When a pattern maker is given a drawing, he has to imagine the casting before him, and build something that will produce it; it may be called a pattern, but often it is really not a pattern of what is wanted, because of the complexity of the casting; it is sometimes all core-boxes and no pattern, and here is where the responsibility comes in, and will, I think, explain why the pattern shop is often the birth-place of mistakes.

Of course, mistakes ought not to occur; but as long as pattern makers are fallible, they will occur sometimes, though the utmost precaution be taken. I am always suspicious of the man that never makes mistakes; he is not to be trusted; but I have no sympathy for those careless pattern makers who are constantly making blunders, and who think when their patterns come within an eighth of an inch it is near enough.

From the nature of the trade of machinery pattern

making, there is more danger of errors being made in that branch of machinery building than others, and the careful, industrious, workman, who seldom makes an error, is worthy of consideration when he does happen to be caught, for such a man usually feels bad enough over his mistakes, without having anyone make him feel worse.

Owing to the advance made in mechanical arts, pattern making is becoming one of the most important branches in machinery building. It is often underrated by a class of machinists who think that because a pattern maker is not called upon to work in iron, and to one-hundredth or one-thousandth part of an inch, that there is not much in pattern making; and yet the pattern maker is as much of a machinist, in reality, as those generally known as such.

. The onward march of improvements in machinery demands that the pattern maker must keep right up abreast with the times, although he is considered "a necessary evil" among manufacturers.

There is a great deal of machinery now constructed, the coring of which is so complicated that it taxes the ingenuity of both pattern maker and moulder to know how it can be made at all—the winding passages and secret chambers that are wanted in some castings, are worse than those we read about in books. The old fashioned idea of bolting on an arm here, and screwing on a bracket there, are fast dying out. The modern plan is to make a machine with as few pieces as possible, thus making the pattern more difficult to build.

There are many patterns that require little or no

knowledge of pattern making to make, but I would not advise anyone, because he has made a few such patterns, to pose as a pattern maker; there are those who do. I have had some experience with them, and hope always to be delivered from such. They are a worry to any foreman—he is in constant fear that with all his watching, the would-be pattern maker will make some serious blunder that will cost the firm a considerable sum of money—for a mistake in the pattern means a mistake in the casting, and as an old employer of mine used to say: "Cast iron mistakes are rather serious things."

The fact that there are so many different ways of moulding, gives a great field for study for the pattern maker, as to the best way of making a pattern; but whenever a complicated piece of work is to be done, the moulder should be consulted, and I do not think that the pattern maker will lose any of his ideas by consulting with his brother, the moulder, and while the practical parts of the two trades are as unlike as possible, yet there is a connection between the moulder and the pattern maker that is inseparable. If discussion is necessary, let it be carried on intelligently, each respecting the other's opinions. Wherever this is done, good is sure to result, and the chances are that the best way of doing a job will be arrived at. There are those who are so eager to advance their own ideas and have them carried out, that they are unwilling to consider those of others; such persons are not likely to be very profitable to any concern, for they think more of airing their own genius than of arriving at any results that might be of practical value.

THE PATTERN SHOP.

ITS POSITION, SIZE, AND REQUIREMENTS.

THE question has often occurred to me why pattern shops are located on the upper floor of a building, as they usually are. The foundation for fast running machinery is anything but good on an upper floor, besides being very inconvenient for getting patterns up and down. It also is risky business turning a large pattern in a lathe whose only foundation is an upper floor that springs with every motion of the machinery; the chances that pattern makers will take when turning a large pattern under these conditions are great. The ground floor is a much better location for a pattern shop.

Large face lathes for turning large diameters cannot be too rigid, but ofttimes the trembling of the face plate is caused by too small an arbor, or the bearings may be too close together.

Plenty of room and light are two essentials that are generally lost sight of in arranging for a pattern shop. This shop is sometimes called the pattern room, and I suppose it is thus named, from having, as is often the case, such a small space set apart for that purpose, that it has scarcely deserved the name of shop.

The nature of the trade, in a large measure, determines the size pattern shop a firm requires. A firm of large dimensions making specialties does not need such a large pattern shop, as a smaller one that builds engines and general machinery. It is more to this latter class of manufactories, employing about twelve or fourteen pattern makers, that reference is made.

Go into a number of manufacturing concerns, and in nine out of ten, it will be found that the pattern makers are working so closely together as to prevent them from getting around their work in a proper manner; and it is surprising how a job may be impeded for lack of room to build it.

When it happens that there is a run of large work, then it is that the oft repeated expression is heard "We ought to have a larger pattern shop." It is granted that shops are situated on such valuable property sometimes, that a limited space only can be allotted to each department; but this does not do away with the fact mentioned. In these days of sharp competition, the firms that are not cramped for room are the successful competitors in the machinery business.

A pattern shop about 75 ft. x 50 would be a convenient size for working the number of men named.

The machines should not be located all over the shop, but at one end within a reasonable working distance of each other.

Among the requirements of such a shop would be a face lathe for turning large patterns, 30" lathe with bed about 18 ft. long, 16" lathe for small work, combination circular saw table, plain saw table, with saw about 12" diameter, band saw, jig saw, surface planer, Daniel's planer, two or three Fox Trimmers, and about six dozen (rather more than less) of assorted clamps. I mention

Fig. 1.

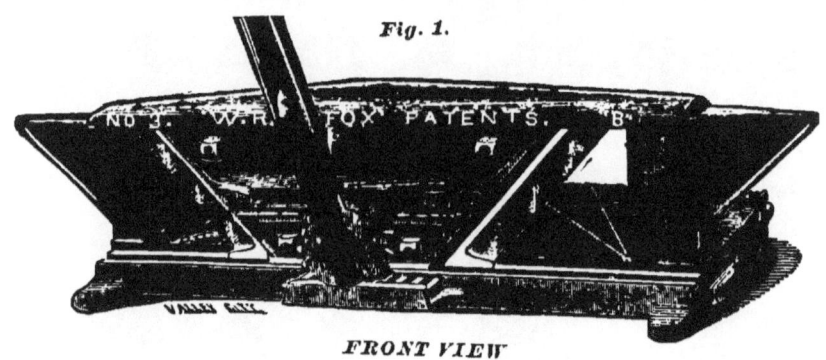

FRONT VIEW

Fig. 2.

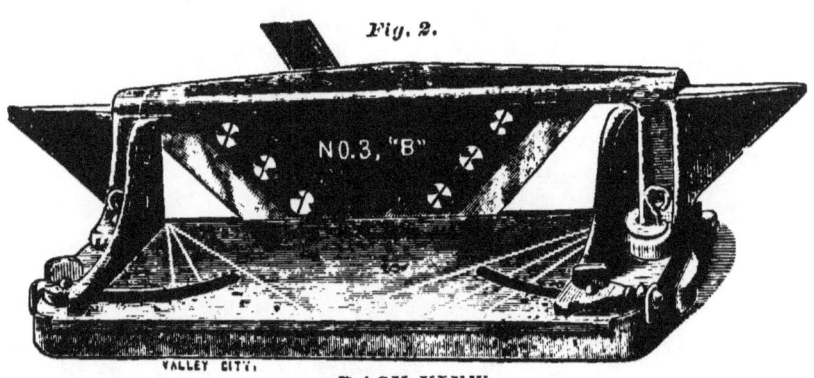

BACK VIEW

6 in. STROKE: 8 in. RISE: BED 6 x 15 inches: Weight 25 lbs:
For SMALL PATTERN WORK.

FOX'S TRIMMER.

this smaller item of clamps in order to insure a plentiful supply. Much time is frequently lost by men waiting on each other for clamps.

The Daniel's Planer is a machine that no pattern shop of any pretentions should be without. For surfacing stuff for pattern makers this machine has no equal, especially when knives are kept sharp, and a good supply should always be on hand.

The Trimmer mentioned is also a very valuable addition to the pattern shop, in fact, it has come to be a standard tool, and the shop that is without one is away behind and had better hurry up and get at least one.

I believe the success of this machine and its being adopted so generally in pattern making, is due to the fact that it was invented by a pattern maker who designed it at the time for pattern making. The Trimmer has certainly done away with a great deal of the paring that used to done with a chisel, and which was exceedingly laborious, as most of my readers know, especially when cutting end-way of the grain. For building up segment work the Trimmer has become almost an indispensable tool, and will cut as straight and as clean as it is possible to cut wood. Figs. 1 and 2 are two views of the smallest size Trimmer made by the Fox Machine Co., Grand Rapids, Mich. The illustrations will give the necessary explanation and will be readily understood. The Company make several sizes and the most fastidious "wood butcher" can be suited.

It is not necessary to go into the details of pattern shop requirements, but there is a mechanical paper published that ought to be considered a requirement for pattern makers, and that is the *American Machinist*. There is no doubt but that this is the best and cheapest technical educator we have, for it contains more practical ideas for doing work in all the branches of the machinery business than most papers.

MARKING AND RECORDING PATTERNS.

THE practice of fixing a mark or symbol on a pattern to distinguish it from others, is an excellent one. The pattern department of any firm cannot afford to disregard the marking and recording of its patterns. In many places a large stock is accumulated, regardless of any system; the man who looks after them calls it "Red Tape" to mark patterns and record them, and says, "I know where to find any pattern without any such nonsense;" at the same time they may be piled together like a lot of kindling-wood. What this rule of thumb individual says about knowing where to find any pattern may be true, but should any unforeseen circumstance remove him, who is to find the patterns and know about each, then?

The disadvantage that such a firm labors under through not adopting some system of marking and recording is great. To those who have no system I would recommend the following :—

Fix a raised letter and number on the pattern, so that it shall appear in the casting. The letter is to designate the class of machinery, or it may be used for a certain machine; the number, to distinguish one part from another. The mark that each pattern gets should also be put on the drawings. This is not generally done, but I think if it were, it would greatly facilitate the work in the machine shop. The method which I worked for many years is shown by the sample entry of patterns given on page 11. The column "pattern at" will be found very useful to those sending out their patterns; it is intended to show where the patterns are. In connection with this column an index is made showing the names of firms with whom business is done. Each firm is given a number, as shown; when a pattern is sent out, the number corresponding to the firm it is sent to is marked with lead pencil in the column, "pattern at," and opposite, the pattern that is sent out; when it is returned the lead pencil mark is rubbed out, showing that it has been returned and stored in its place.

It often happens that in a set of patterns for a certain machine, there are those that will do for other machines; in such cases an entry should be made in the schedule of each machine that this piece will suit, giving the same letter and number of the pattern.

There is always a large number of miscellaneous patterns that cannot be so well classified, yet many of them are often used and need marking; these may be given a symbol and entered under the head of "miscellaneous."

There are also some rough patterns made, the kind that is generally "wanted to be cast to-day." All moulders are acquainted with this kind. It is no use recording such patterns, as they are seldom used the second time; in fact, I think the best way to deal with this class is to break them up.

With large manufacturers carrying patterns for a number of different machines and many classes of machinery, the question may arise what to do when the alphabet is exhausted. When that happens, two letters can be used to designate a machine or a class, commencing with *AB*–

Sample Entry of Patterns.
18″ AND 20″ CORLISS ENGINES.
A.

Symbol	Name of Part.	Castings Wanted.	Weight.	Pattern At	Remarks.
	CAST IRON.				Pattern arranged so that it may be set for various strokes.
A. 1	Frame for 18″ and 20″ Engines, - - - -	1			
" 2	18″ Cylinder, - - - -	1			" " "
" 3	18″ Cylinder Head, - -	1		1	Arranged to be set to various strokes.
" 4	20″ Cylinder, - - - -	1			
Etc.					
	BRASS.				
A. 41	Key Heads for Motion Rods, etc., - - - -	10			
" 42	Connecting Rod Brass, -	1		2	Cross Head End.
" 43	" " " -	1			Crank Pin End.
Etc.					
	CAST STEEL.				
A. 51	Cross Head, - - - -	1		3	
" 52	Eccentric Rod Nut, - -	2			
" 53	Cross Head Nut, - - -	1			

INDEX OF FIRMS AND THEIR NUMBERS.
1. EDDY FOUNDRY CO. (*Iron*), CHICAGO.
2. THOMAS BROS. M'F'G. CO. (*Brass*), CHICAGO.
3. EUREKA CAST-STEEL CO. (*Steel*), CHESTER, PA.

AC–AD, etc., until through the alphabet again; then begin with *BC–BD–BE*, and so on; thus, it will be seen that it is possible to have a large combination of distinct and separate classes without confusion.

The advantage of recording and marking patterns is that it facilitates ordering the castings and helps to prevent confusion in the foundry. When the order for castings is written out (as it always should be) for the foundry, the mark corresponding to that on the pattern is put on the order so that the moulder and the pattern maker cannot misunderstand each other by naming things differently. Again, when a pattern has a particular mark, every loose piece (and sometimes there are a great many) belonging to the pattern, and also the coreboxes, can be stamped with a mark corresponding to the pattern. The benefit of this is apparent. There is often much trouble caused by not knowing where a certain loose piece belongs, and castings are frequently made minus a piece just because the moulder did not know that it belonged to the pattern; but if every piece is marked as I have said, it leaves no excuse for such omissions. The raised letters that are nailed on the pattern help greatly in checking the castings when received. Especially is the marking of patterns necessary for this, as gentlemen of the quill profession, who generally check the goods, are not usually acquainted with the names of the parts of machinery. Also by this method the finding of patterns is rendered easy even to a stranger; that is if the shelves where patterns are stored are marked with the letter corresponding to the class.

Every firm, large or small, should have some such system as I have described.

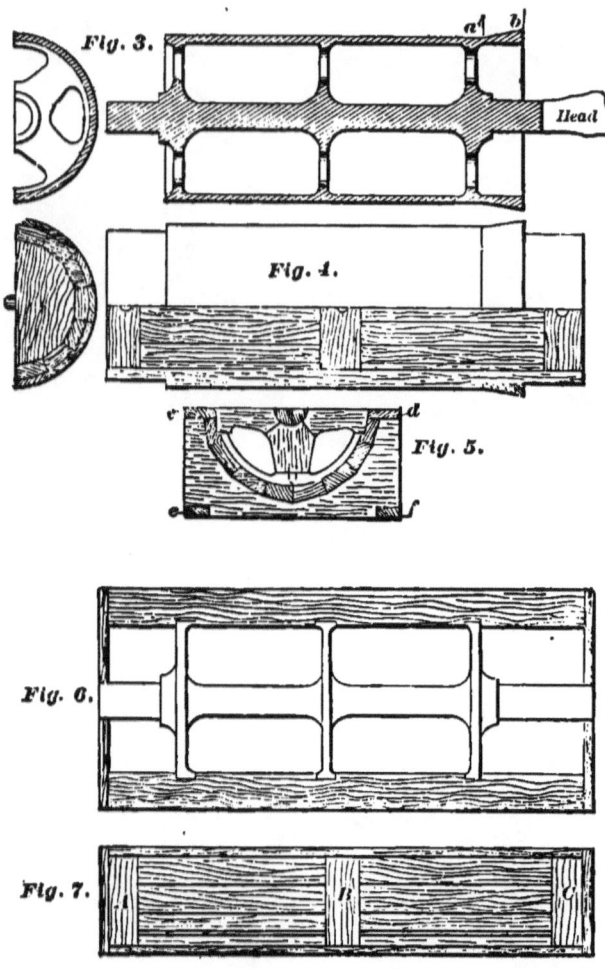

PRINTING-PRESS CYLINDERS.

PRINTING-PRESS CYLINDERS.

SOME printing-press cylinders have the ends bored out and a short shaft pressed into each end, while others are made with the shaft and cylinder cast in one. Fig. 3 is a section and end view of the latter.

A great deal of trouble is often experienced in getting perfect castings for these cylinders, and to insure good castings they are cast on end in dry sand, or at least they should be; but in spite of all the preventives used against blowholes, dirt, etc., a cylinder will sometimes reveal defects when the first cut is being taken off in the lathe.

Though cast on end, the cylinder is moulded on its side, so that the pattern is made in halves in the ordinary way, as shown in Fig. 4.

As an extra precaution against defects an additional piece five or six inches long is cast on the end, as shown in Fig. 3 from a to b., the pattern therefore should be made that much longer.

The extra length will receive the impurities of the metal which rise to the top when pouring. It is made thicker on the ends, so that it shall form a head which will exert a pressure, thus helping to produce a clean and sound casting.

Figs. 5, 6, and 7 are three views of the core-box. The end view, Fig. 5, shows it to be built up with staves, which are nailed to three crosspieces, *A*, *B*, *C*. The box is strengthened by running four strips, c, d, e, f, lengthwise on top and bottom of the box, fastening them to the crosspieces.

Two of the arms in each set are let in about $3/8''$ on the inside to keep them from being rammed out of place, but a dowel pin is put in each of the arms that go in the bottom of box.

This is a very plain and simple job in pattern making and needs no further comment.

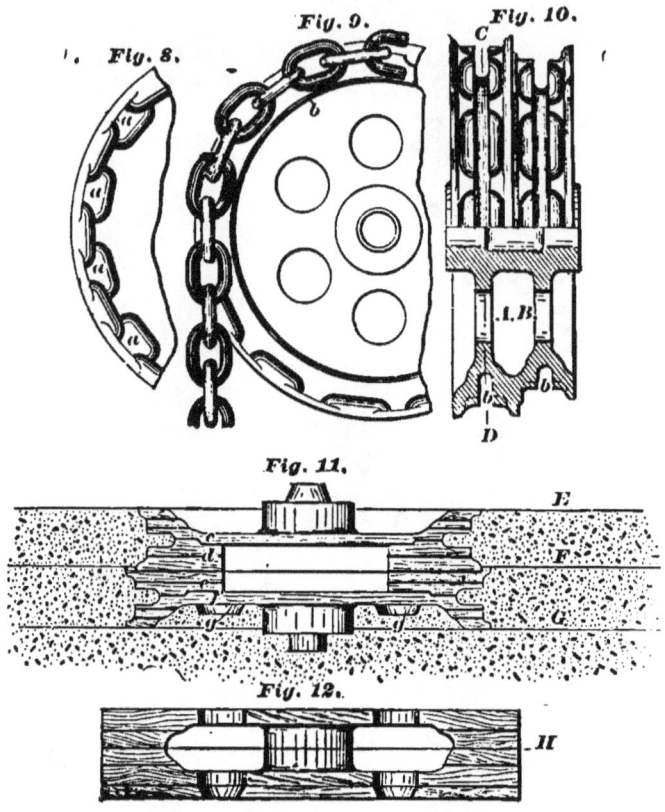

DIFFERENTIAL CHAIN WHEELS.

DIFFERENTIAL CHAIN PULLEYS.

WHEN the groove in a chain wheel or pulley is made to fit the links of a chain it is sure to be an expensive pattern, especially when made double, like those used in Weston's Differential Pulley Blocks.

Figs. 9 and 10 are two views of one of this kind of pulley. Fig. 10 shows a half section and the pockets for the chain. Fig. 9 is a section through the groove C, D, of the large pulley, which has one more pocket than the smaller one. This view also shows how the chain fits into the pockets.

It may not be out of place here to make a few remarks about this celebrated Differential Pulley Block that has so revolutionized the lifting of heavy weights, and for which this kind of pulley was used, in fact, this pulley was the main feature of the patent. T. A. Weston was in this country when he conceived the idea that led up to the Differential Pulley Block.

While Mr. Weston was at Buffalo, witnessing attempts to raise a vessel that had gone down off that city, the thought occurred to him that the necessary power could be obtained from the Chinese windlass, the rope of which winds on two unequal diameters, that is, one half the length of the barrel is larger in diameter than the other. This is practically what this pulley block is developed from.

After much scheming, Weston returned to England and called on numerous engineering establishments, submitting his drawings, but he could find none that would take hold and experiment on his block. Finally he called into a small job shop where the proprietors themselves were working men, paying the extravagant rent of ten shillings a week, and employing six men. That firm has grown since then and employs as many thousands now. I refer to Tangye Bros., Birmingham.

These brothers labored hard to make the block work, and experienced many unexpected difficulties, and when they had perfected it and made it a commercial success, a new difficulty presented itself in the shape of a law-suit in which the Tangyes won. It is pretty hard to conceive of any taller swearing than was practised by the would-be infringers in this case, but I will return to the pattern of the pulley.

They are sometimes made so that all the links of the chain fit into the pockets, but this is an unnecessary expense. The links of the chain that set edgewise in the pulley do not need to fit into pockets like those shown at a, in Fig. 8. If the grooves bb, in Fig. 10, be turned deep enough to clear these links and pockets made for the other links to set in, it will be sufficient to catch the chain, and will work better than otherwise.

The groove is sometimes formed in a core-box, and a print put on the periphery of the pattern, thus making fewer partings in the patterns as well as the mould; but a much cleaner and better casting can be obtained from a pattern with the groove for the chain cut in it.

Fig. 11 is the section of the pattern and shows how it

is made. The mould is also represented with the cope lifted off, the partings being at *E, F, G*. The pattern is built up with segments and made in four parts, *c, d, e, f*. As will be seen, the casting is cored out at *A. B.*, in Fig. 10.

Fig. 12 is a section of the core-box for this core, and is parted at *H*. The core sets into the round prints *g. g.*; but there are no cope prints, for the reason that it is not easy to close the cope over the six round projecting cores. In the absence of these cope prints the moulder will need to take care that the cope bears on the top of these projecting cores enough to prevent the iron from running in the vent holes of the core, when pouring.

This lightening core can be made in halves or whole, just as the core-maker chooses.

A HANDY TOOL EOR LAYING OUT HEXAGON NUTS.

FIG. 13 illustrates a tool for laying out hexagon nuts, and is very handy to pattern makers; the section of it is shown at *A*. The upper part, which is a light steel blade, is screwed on the lower part, which is made of hard wood and is used in the following manner. After turning pattern to long diameter of nut, place the tool on pattern like a center square, move it round and mark off sides—keeping the two under edges in contact with circle—this is better and quicker than dividing off with compasses and then marking sides.

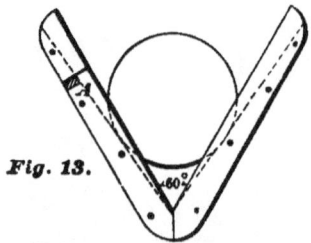

Fig. 13.

Tool for laying out Hexagon Nuts

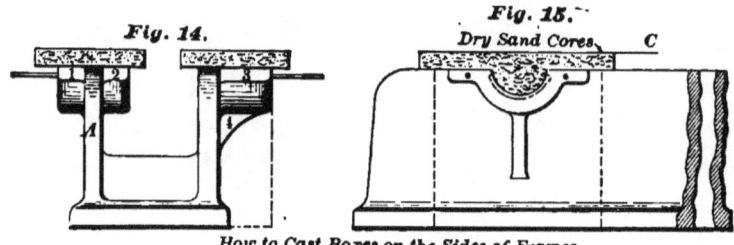

Fig. 14.   Fig. 15. Dry Sand Cores

How to Cast Boxes on the Sides of Frames.

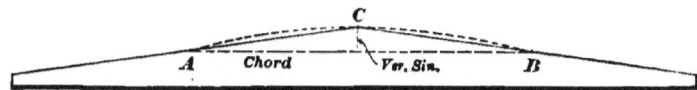

How to Strike a Curve when the Centre is Inaccessible
Fig. 16.

HOW TO CAST JOURNAL BOXES ON FRAMES.

THE part of a frame shown in Figs. 14 and 15 is one of those jobs that at first looks a little troublesome for moulding, and yet, upon examination the trouble vanishes. The two views show part of a frame with two boxes cast on the sides. The shape of this frame is such as to necessitate casting the boxes down; it will be seen that there is not enough room to draw in the boxes, the sides, A and B, not being thick enough to allow it. This difficulty may be overcome by making the pattern as shown. The boxes are loose and located on the sides of pattern with loose dowel pins that can be pulled out while ramming up; two cores are made and dried for the boxes, and rammed up with the pattern to C, after which the cores are taken out, and the sides of boxes, 1, 2, 3 and the bracket, 4, are drawn. The cores for boxes are then replaced and covered over with sand, the flask rammed up and rolled over. There are other ways of making this pattern; a core print might have been put on the pattern, as shown by dotted lines, and a core-box made with box pattern in it; but the above way of doing it makes a cleaner job. This plan is adopted on many jobs where there is not room enough to draw in loose pieces.

HOW TO STRIKE AN ARC BY THE AID OF THREE POINTS.

It is sometimes required to lay off an arc, the radius of which is given; but the radius may happen to be so great as to render it inconvenient to locate a center to strike the arc from, or the center may be inaccessible from various reasons; under these conditions the question arises what to do to get the arc.

The following is by no means new, yet it is so little understood by pattern makers generally, that I think it worth while presenting. The all-important, in this problem, is to know how to get the point C, or the versed sine of arc in Fig. 16; this must be calculated before anything can be done towards striking the arc. Let us suppose the line AB to be , say 4 ft., and to represent the chord of an arc whose radius is 20 ft.; it is required to strike this arc without using the center. By using the following formula the desired point can be obtained: $v = R - \sqrt{R^2 - c^2}$. This formula need not scare anyone who is not familiar with algebraical expressions; it is very simple, let us examine it. v is the versed sine, R, the radius, c, the semi-chord. v is what we want to get. The formula means that the square of half the chord, which is 4 ft., must be deducted from the square of the radius, which is 400 ft., this will give 396 ft.; then extract the square root ($\sqrt{}$) of 396 ft. which is about

19.9: the formula is now reduced to v=R—19.9. which means that 19.9 must be deducted from R, the radius, 20 ft., this leaves $\frac{1}{10}$ of a foot; $\frac{1}{10}$ of a foot is $1\frac{3}{16}''$+which is the required versed sine.

Having obtained the versed sine of this arc, it is an easy matter now to strike the arc—it is done by cutting a piece of wood to an angle, two sides of which run from point *C* and through *A B*; drive a wire nail at each point of *A* and *B* in the piece on which the arc is to be struck. It will be readily seen now, that, keeping the sides of angle against *A B*, and moving it right and left, the arc can be traced by following with a lead pencil the point *C*. The principle of this is the same as many pattern makers are familiar with—that of using a square for a templet when working out a half-circle core-box.

KEY HEADS FOR MOTION RODS.

AN EASY WAY TO LESSEN THE COST OF THEIR PRODUCTION.

The cost of getting out brass key heads for motion rods may be considerably reduced in the machine shop by the pattern maker doing a little scheming and the brass moulder exercising care in moulding.

Figs. 17 and 18 are two views of a key head, with block A in place. A dry piece of cherry or mahogany should be selected and the pattern made as shown in Figs. 19 and 20; it should be in halves, $B\ C$ being the parting line. $D\ E$, Fig. 20, are core prints which carry the core horizontally. After the core is located a steel key is set in the mould into the print a, the cope print b bringing the key upright when cope is being closed. The box for the core is shown in Figs. 21 and 22 and is doweled together at $c\ d$, Fig 21; a key passes through this core-box, which makes a groove in the core to receive the steel key.

It is the intention, when the castings of these heads are being fitted up, to file the round ends, while the sides are to be finished in the machine, therefore the stock allowed for machine finish must stop off at $f\ g$.

There need be no fear of chilling the sides of the holes through casting steel keys in these heads; the effect is rather the reverse of what would happen in cast iron, for it is well known that if the same thing were to be done

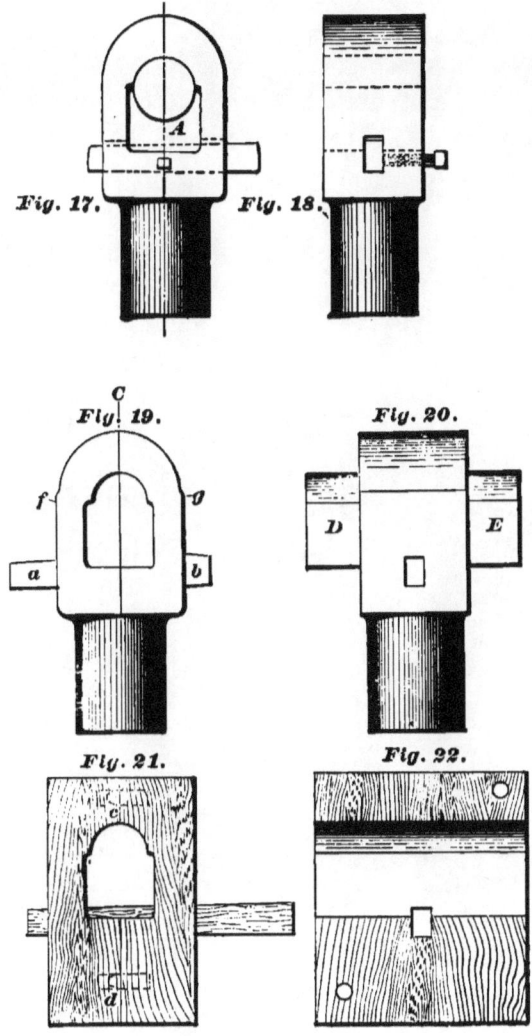

KEY HEADS.

in cast iron, the sides of the holes would be so hard that it would be almost impossible to dress them out with a file. A number of keys should be made expressly for casting into the heads, and they ought to be nicely finished, having about $\frac{1}{64}''$ taper sideways; this will enable them to be easily driven out.

I have seen much time wasted in the machine shop making these key heads, such as drilling and chipping out the key hole from the solid, and finishing the round end in a shaper. This led me to devise the above simple way of casting them, which has proved a great saving.

When making large quantities of these castings, a wood pattern will necessarily get broken and badly marked with the moulder's vent wire, so that under such circumstances, a metal pattern and core-box would be more serviceable. I would therefore propose that the standard pattern be made of aluminum and the core-box of cast iron. Aluminum is just the metal that is wanted for such small patterns, because it possesses the two necessary and important elements most desirable for patterns, strength and lightness in weight; a very nice surface can also be made on this metal, which is also the thing needed.

Moulders do not like a heavy pattern, for the reason that it is not so easily drawn as a light one.

ELBOW AND TEE PIPES.

A QUICK METHOD FOR TURNING THE PATTERNS AND CORE-BOXES IN THE LATHE.

MAKING patterns for elbow and tee pipes, if made the right way, is comparatively simple, because nearly all the work can be done in the lathe. For turning out a large number of castings the elbow pattern should be constructed as shown in Fig. 23, so that two elbows can be moulded together. A ring is turned like Fig. 24, the section of it being a half circle, the same size as the pipe; this ring is cut in quarters as shown, and the four pieces used to make quarter turns for the elbows. In Fig. 23, the two spicket ends A and B, and the sockets, with core prints, are turned on one stick and cut off; the stick should be sawed off long enough to permit of tongues being turned on A and B, and the sockets, for fastening them to the elbows; dowel pins should be arranged to come one in each socket, and in the print between A and B, as marked. Fig. 25 represents the core-box, and, like the pattern, most of the work can be done in the lathe and the parts joined together, as shown; a piece screwed on the back will hold the parts together. The core is generally made in halves, so that a full box will not be needed, and therefore only two quarters are used for the turns. In some cases these two quarters that are left may be used to turn the socket ends

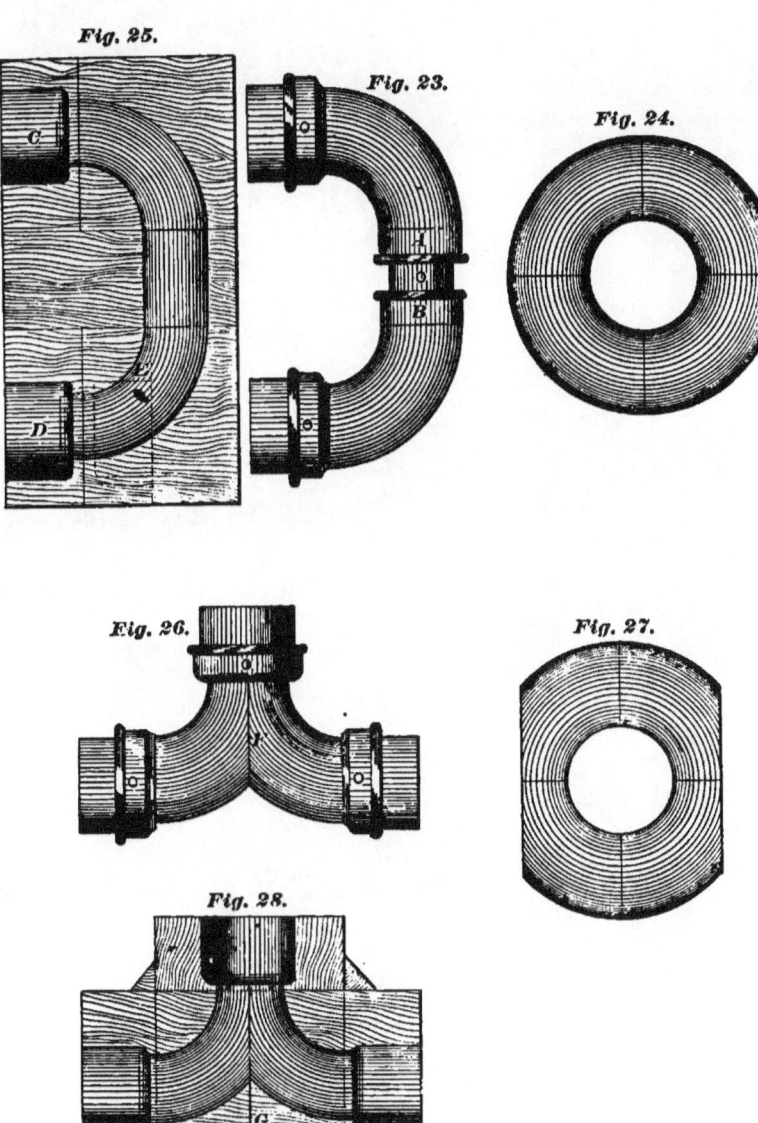

ELBOWS AND TEE PIPES.

C and *D*. The dotted lines show that in this case; the pieces left over cannot be used without cutting out the corner *E* and inserting a piece; in a small elbow it would not be worth while to do this, but for a larger pattern it would probably pay.

Figs. 26, 27, and 28 show the manner of making a pattern for a tee pipe, and will need but little explanation, as it is made much in the same way as the elbow, all the parts being turned and fitted together. In addition to being quartered, the sides of the ring in Fig. 27 are cut off and the parts joined together at *F*, Fig. 26. The parts for the core-box are also cut this way. Before gluing the joint *F* together it should be sized with glue, after which the joint will be very strong when put together; the socket will also help to hold the pattern together at this point. This form of tee is to be preferred to the right angle ones, because of its extra strength and the gradual merging of the passages into each other, which renders the flow of water, etc., easy.

SLIDE VALVE CYLINDERS.

SLIDE valve cylinders are made in a great variety of forms. I have chosen and represented here a well-known type, of which Figs. 30, 31, and 32 are three views. Fig. 31 is a cross section through the steam chest and exhaust port, and Fig. 32 is a section through the steam port.

The way of making the pattern for this cylinder depends largely upon the size of it; if the diameter is to be, say less than 12", the body of the cylinder may be built up solid, but when above that size it would be better to build the pattern with staves, as shown in Figs. 33 and 34. But one-half of the pattern is shown, which will be all that is needed for explanation. An extra thickness is glued on each stave large enough for turning the body of the cylinder to the required diameter, thus allowing the prints and the cylinder to be in one, which is far better than fastening on the prints. The flanges are made thick enough to turn the fillet on the back of them. They should be got out, as shown in Fig. 35; this will prevent the shrinkage of the wood from affecting the flanges to any great extent.

When the body of the cylinder is built up and turned, the steam chest is made and fitted on, as seen in Figs. 36 and 37. The pieces for the exhaust passage A, and those to form a thickness over the ports, are also shown

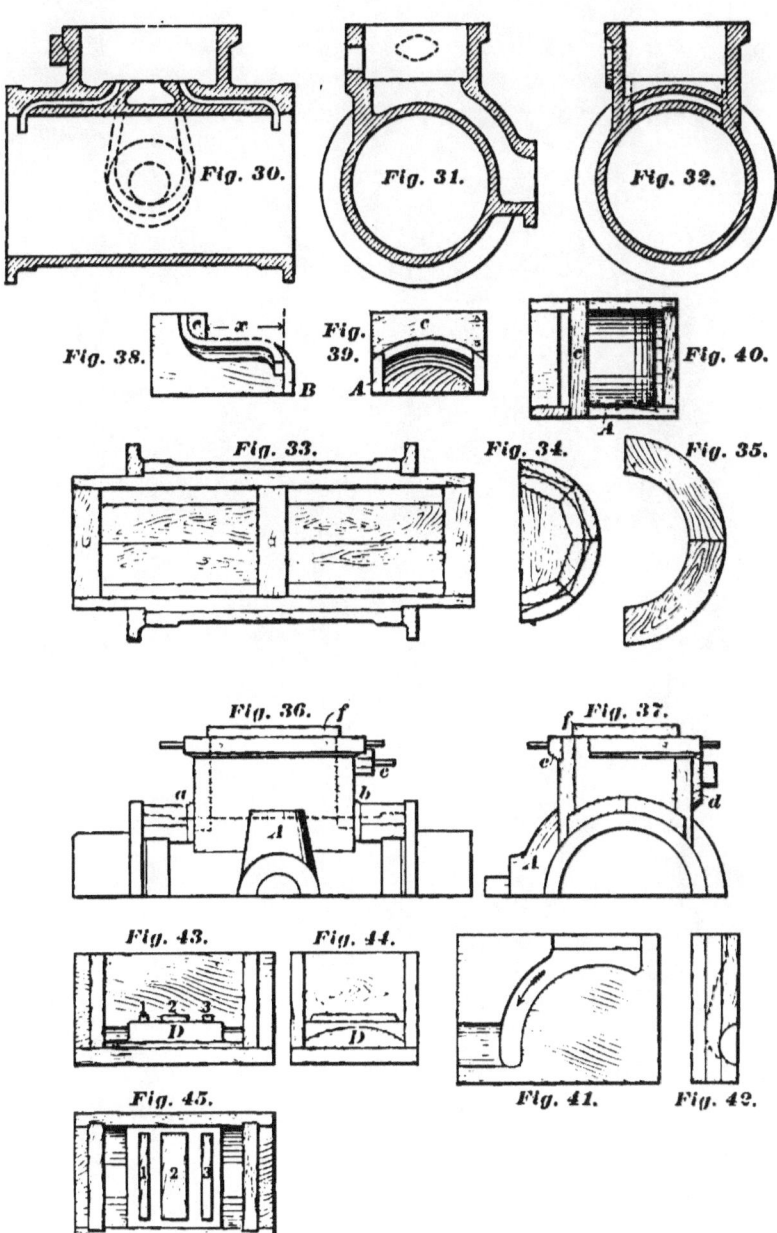

SLIDE VALVE CYLINDERS.

in place. The fillets at *a b*, Fig. 36, are cut out of a thickness that is inserted between the end of the steam chest and port pieces; there is no danger of the fillets coming out when made in this way.

The strips on the steam chest, which give an extra thickness of metal for the studs, are loose and put on with long dowel pins, so that they can be drawn separately, also the valve stem stuffing box, *c*, and the facing, *d*, around the steam opening are loose. The strips are shown at *c*, Fig 37, let in about $\frac{3}{16}''$ around the sides of the steam chest; this is done to prevent them being rammed out of place after the dowel pins have been taken out. The core print *f* should be doweled on, because if it is made fast it is very probable that the moulder will tear it off, for it is a great convenience for him to have this print to place back in the bottom of his mould while dressing it up.

Figs. 38, 39 and 40 represent the core-box for the steam port. Fig. 38 is a side view with the side *A* taken away. The core is swept off on the outside for the length of *x*, but the piece *c* is made to finish the outside, because the core changes from a circular into a a straight part, just where it is entering the steam chest. Fig. 39 is an end view of this box, with the end *B*, in Fig. 38, taken away. Fig. 40 is a plan view and will be understood from Figs. 38 and 39.

The exhaust port core-box is made in halves; one-half is shown in Figs. 41 and 42; the other half is, of course, like this, except that it is made the opposite hand, so that the two halves shall fit when put together. The dotted line in the end view, in Fig. 42, shows how the passage is widened to maintain the same area throughout; that is, it is cut down along the part in the direction of arrow in Fig. 41.

The steam chest core-box is made as seen in Figs. 43, 44, and 45; the side and end are taken off in Figs. 43 and 44, to show the inside; the piece D that forms the valve face is screwed on the bottom, and the sides fit over it; this core is the first that is set in the mould, then the exhaust and steam port cores are set into 1, 2, and 3, Fig. 45.

Univ.

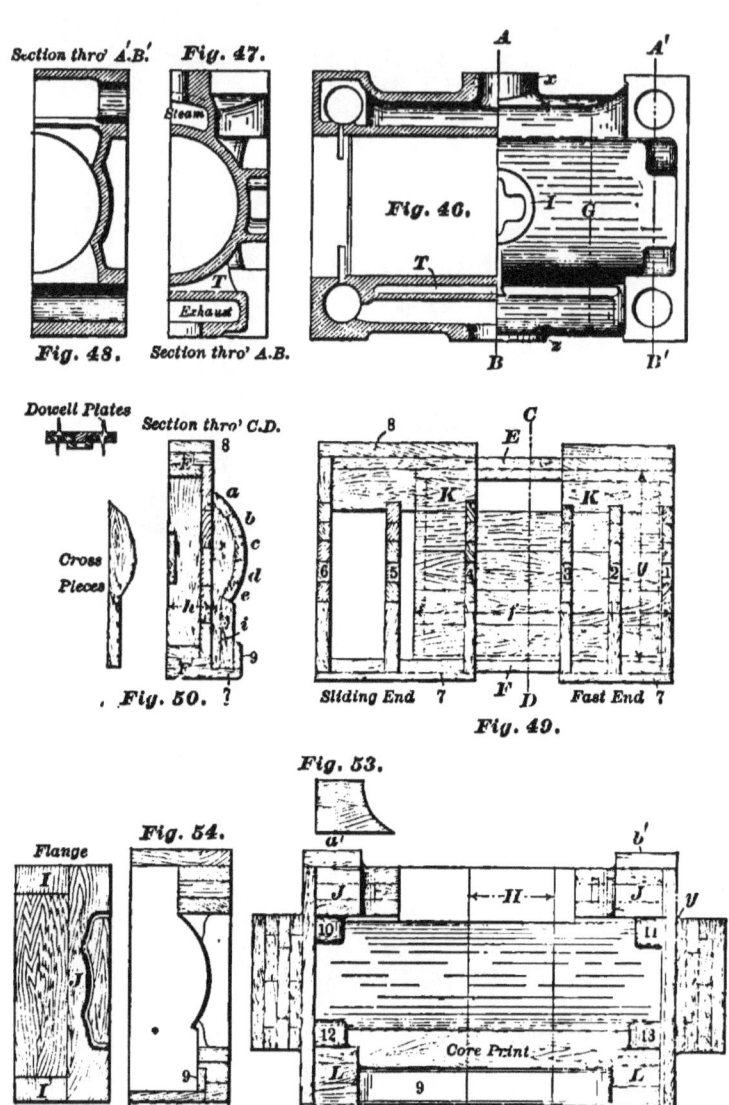

CORLISS CYLINDERS.

CORLISS CYLINDERS.

SHOWING HOW TO CONSTRUCT PATTERNS AND CORE-BOXES
WHICH CAN BE CHANGED AT SHORT NOTICE
FOR DIFFERENT STROKE ENGINES.

It is often required in shops building engines that a cylinder pattern be so constructed that it will serve for engines of different strokes. To illustrate one way of making a pattern like this I have chosen a Corliss Cylinder. Fig. 46 shows the wrist plate side and half section of cylinder. Fig. 47 is a section through $A\,B$, and Fig. 48 is a section through $A'\,B'$. As both halves of pattern are almost alike, it will only be necessary to deal with one half.

Experience has taught me that $\frac{1}{12}''$ to the foot is enough to allow for shrinkage on a job of this kind. Use first quality dry lumber; it will pay to take some pains in selecting stuff for this pattern; for, if the lumber is not thoroughly dry, before the pattern is completed, it will be found that dimensions are scant.

By referring to Fig 49, it will be seen, that to make a pattern for different stroke engines, I have arranged it to slide like a telescope. Fig. 49 is shown pulled apart, and illustrates how the pattern is built. Fig. 50 is a section of Fig. 49 through $C\,D$; the lagging $a\,b\,c\,d\,e$, in Fig. 50, is taken off in Fig. 49 to facilitate explanation. An inner box is first made, the length, width, and depth of f g h in Figs. 49 and 50, and doweled together with iron dowel plates like sketch. Build out each side of this inner box for about two-fifths of its length to width G, in Fig. 46, then fasten on strongly three cross pieces numbered 1, 2, 3. For future reference I will name this the "Fast End."

In Figs. 49 and 50, the outside piece marked 7 should be made wide enough for the exhaust passage in Fig. 47. Notice that in making the inner core-box the board K is cut in two so that one piece may form a part of the sliding end. On the sides of box, glue and screw two guides marked E F in Figs. 49 and 50; on each side of these guides, fit pieces that shall slide over them, and secure them to the two outside pieces, 7 and 8. This will make the sliding end the same width as fast end. Now fasten the cross pieces, 4, 5, 6, and piece K, to the two slides, and be careful not to get any of them glued to the box; the lagging a b c d e, in Fig. 50 will hold this sliding end together, and, as the pattern progresses, other parts will make it secure. The piece i forms the core print for coring out the space T, that separates exhaust passage from body of cylinder. (See Figs. 46 and 47).

In Fig. 49 we have now a foundation to build on, and have made it the width of G in Fig. 46. At this stage, the length of the pattern, when closed for the shortest cylinder required, should be the length of the cylinder, minus the thickness of two flanges and $1\frac{1}{4}''$ for fillets on back of flanges. Fig. 51 represents the pattern closed, with a filling piece, H, put into it and the framework of Fig. 49 all closed up.

Proceed next to get out the end flanges, valve chambers, port pieces, etc., and build on. Get out the flanges, as shown in Fig. 52, glue on three pieces, I, I, J, $\frac{5}{8}''$ thick; this $\frac{5}{8}''$ thickness is for carving the fillets on backs of flanges, and makes by far the best job. It must be remembered that the cylinder illustrated here is supposed to be a standard pattern, from which a large number of castings may be taken, so that in this case the best way will be the cheapest in the end.

Glue and screw on the end flanges, shown in Fig. 52, and the side pieces, a' b' c' d', Fig. 51, for the valve chambers. Build some blocks together, same shape as shown in Fig. 53, and fasten them in place at $J J$, to form the ends for steam valve chambers. For very large cylinders these pieces had better be boxed. For the ends of exhaust valve chambers, $L L$, make plain square blocks; now make the exhaust passages the right height, by gluing on piece 9 in Figs. 51 and 54; after this, fit in the pieces 10, 11, 12, 13, Fig. 51, that give a thickness of metal over the steam and exhaust ports. All is now ready for rounding off the corners and side of steam passage, and cutting the fillets.

Check down the two inner edges on ends of valve chambers (See Fig. 46). This is to allow the bonnets to lap over and cover the joint of the black walnut lagging, that these cylinders are generally cased in. I have said that both halves of pattern were almost alike; the difference between them is, the center piece for bolting the wrist plate to is usually on the opposite side of the small bosses that take the indicator pipe; also the valve chambers are made about one inch longer on the cope side of the pattern than on the drag side; this is done to insure solid ends after this extra inch is planed off the casting.

When casting these cylinders, all dirt and flux that is in the mould and metal rise to those four high places and stick there; hence, the necessity of extra stock on the ends of valve chambers.

In turning the eight round core prints for the valve chambers make those in the cope as much shorter than those in the drag, as the extra stock just mentioned; in other words, the core prints should measure the same length from the joint of pattern on both halves. The necessity of this will be seen when we come to make the core-boxes. Give these prints plenty of taper, because it will help the moulder to feel his way when dropping in the cores of valve chambers; it will also help to bring the cores upright when dropping on the cope.

Now fit on the pieces x and z, to take the exhaust and steam flanges (see Fig. 46), and, as these pieces are to be removed for sliding the pattern out and in, they must only be doweled and screwed on. Screw the core prints on these pieces from the back. The center piece, I, Fig. 46, should only be doweled on. Build up the core prints for bore of cylinder, as shown, and turn a flat fillet on them at y, Fig. 51. This will prevent crushing the edge of sand when putting the core in and dropping on the cope. Turn these prints whole and saw in halves with band saw after turning them; for large cylinders it will be better to build these prints with staves. We may now consider the pattern finished.

There are many details which are omitted, because

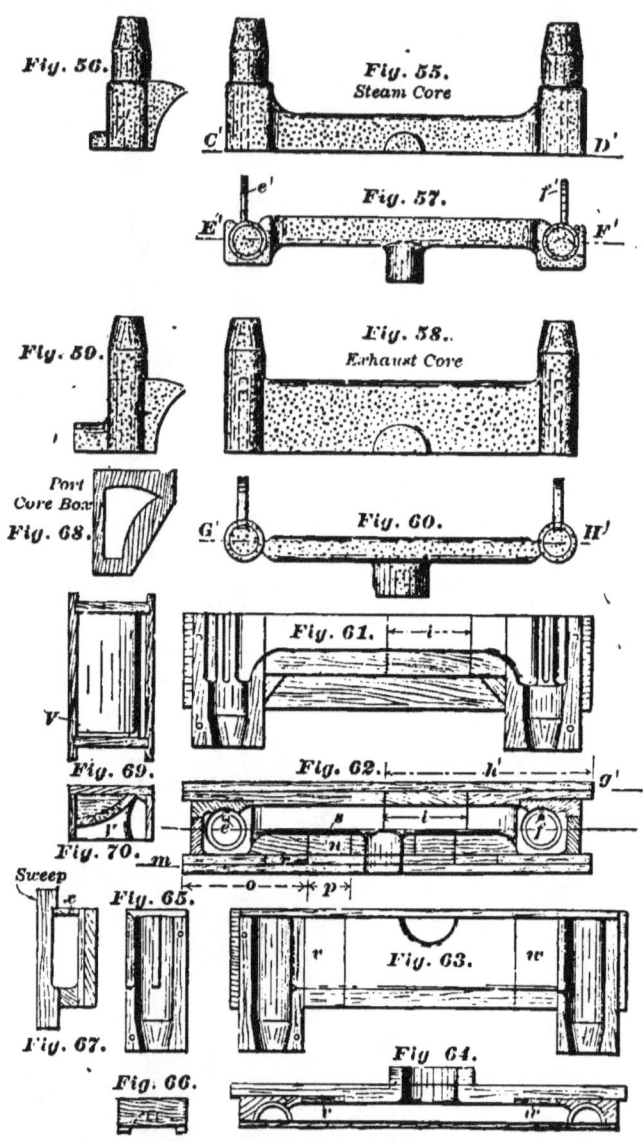

CORLISS CYLINDERS.

PATTERN MAKING. 33

they are such as the pattern maker will naturally run against; so let us leave the pattern and start in on the core-boxes for the steam and exhaust passages, valve chambers, etc.

It will scarcely be necessary to enter into the minor details of making the core-boxes, seeing that I have shown both cores and core-boxes for the steam and exhaust sides of a cylinder.

However, a few explanations may be necessary. Like the cylinder pattern, I have only shown half sections of cores; Figs. 55, 56, 57, are three views of the core for steam passage and valve chambers. A half core-box is all that is needed to make this core, the joint of it being on line C' D', and the joint of core-box on line E' F'.

Figs. 58, 59, and 60 are three views of the core for exhaust passage and valve chambers, also made with a half core-box.

The core-box for the steam core is constructed as shown in Figs. 61 and 62. The port cores, $e'f'$, in Fig. 57, are made separate, and pasted in; the core prints to receive these port cores are seen at $e'f'$ in Fig. 62. Like the pattern, these boxes are made so that the length can be changed. In Fig. 62 the joint, g', for the length of h', is not glued, there being a tongue or guide for this surface to slide in; the piece, i, is loose, which can be taken out, thus allowing the valve chamber part to be slipped toward the center as may be required; on the inlet side of this box, the joint m, for the length of o, and the joint n, for the length of p, is also made to slide, and when the box is to be shortened, the two pieces, r and s, are taken out, which will allow the valve chambers to be slipped toward the center; each side of the inlet is alike. The reason there are more loose pieces on this side of the core-box than the other, is because the inlet must always be located midway between the two valve chambers.

The exhaust core-box of which Figs. 63, 64, 65, 66 and 67 are five views, is made open on one side between the valve chambers, and as this open side is a plain straight surface, it can be swept off with a sweep, as shown in Fig. 67. The valve chambers are round, all the way through, in this core, as will be seen by referring to Fig. 58. Like the core-box for the steam side, two pieces, v, w, Figs. 63, 64, are inserted, for changing the length. Fig. 67 shows a section of the box; the side x, is carried above the center line of valve chambers.

Fig. 60 shows the side of core runs over the center line G, H, hence the necessity of carrying one side of the box above the center of valve chambers.

Figs. 65, 66, are two views of the port halves of valve chambers, and, like the steam core, the port cores are made separate and pasted in.

The core-boxes for the exhaust and steam ports are made like Fig. 68, except that the exhaust port box is made thicker than the steam port box. In small size Corliss Cylinders, whole core-boxes are sometimes made, instead of half-core boxes, but for large cylinders, the cores are better made in halves both for convenience of making and drying.

The box for coring out the space that separates the body of cylinder from the exhaust passage may be made as shown in Figs. 69 and 70. Make the box half the length of the space. (See Figs. 46 and 47 at T). It will be seen that in the center of this space, there is a bridge; to form this, a loose piece, V, Fig. 69, half the thickness of the bridge, is made and fitted in the end of core-box, and as there must be two right and two left hand cores, changing this loose piece to the opposite end will give it. It will be found that two other small cores are needed, one right-hand and one left-hand, to core out between the valve chambers and cylinder on the exhaust side. These small cores are really a part of the core for coring out the space T, in Figs. 46 and 47, and might be made in one core, like the core print shown in Fig. 51, but I think it is easier to make them separately. I have not shown these small boxes, as they will be readily understood without.

Those who make Corliss Cylinders, will no doubt see the advantage of making their patterns to slide the way I have shown, as the pattern can be easily and quickly changed for different lengths without damaging it.

FLY WHEELS.

DIFFERENT STYLES.

MAKING fly wheels is such an every-day occurrence, that it seems almost unnecessary to say anything about it, and yet I think something might be said that might be of service to some one.

These wheels are generally cast in halves and bolted together, except in the case of a fly wheel with square rim, in which case it is generally held together with wrought iron links, shrunk on each side of the rim, as shown at B, in Fig. 71, or sometimes with cotter-bolts, as seen in Fig. 72.

Fig. 71 is part of a square rim fly wheel, and Fig. 73, that of a band wheel. These two wheels are the same diameter and of about the same weight. Let it be supposed each is designed for one size engine; one customer will want a band wheel for his engine and another may want a fly wheel with square rim, and because of this, it is proposed to make the same arm core-box, with changes, do for both wheels. Fig. 74 is a section of a mould for a square rim wheel; it will be seen that the lower and outside part of the mould are formed in green sand, the segment pattern in Fig. 75 being used for that purpose. A step is made in this segment at A, so that in ramming up, the green sand can be swept off level

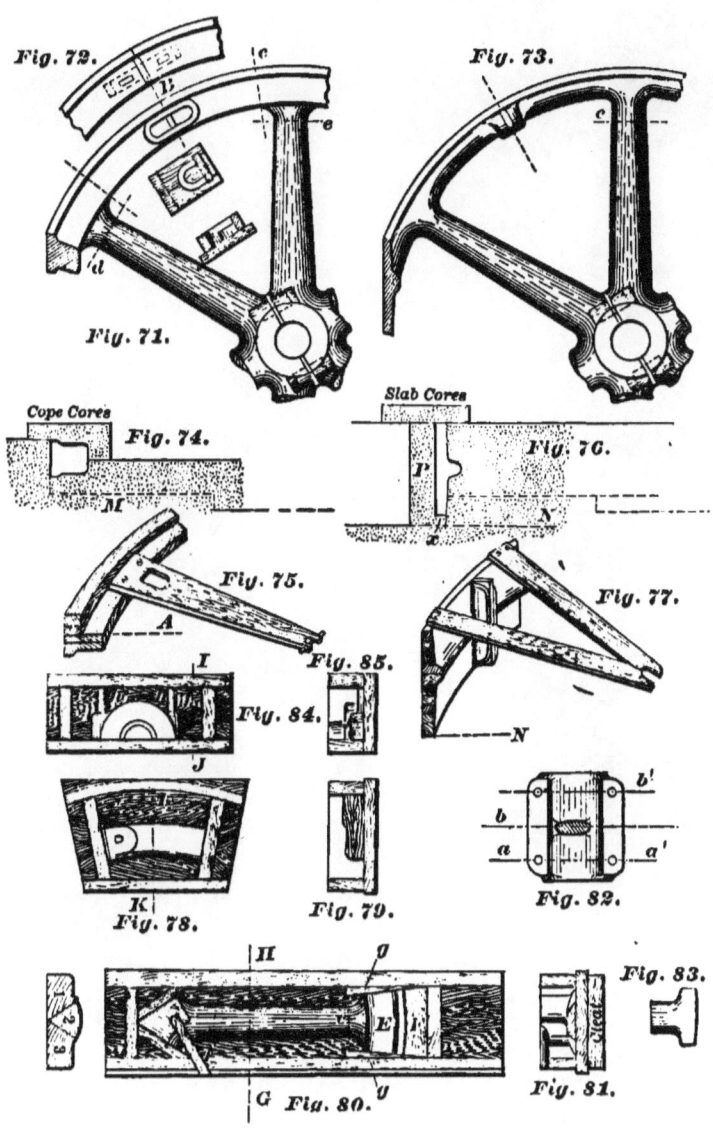

FLY WHEELS.

with it; this makes a level surface on the inside, on which to set the cope cores shown in Fig. 74. A pin is put through the end that fits over the spindle to prevent it from getting away.

Fig. 76 is a section of the mould for the band wheel; the inside of the rim is all rammed up in green sand, using segment in Fig. 77. Two views of the core box for the cope of square rim fly wheels are shown in Figs. 78 and 79, and the length of the box is from B to C, in Fig. 71. D, in Fig. 78, is a print to receive the core for the link, and as there are to be two right and two left hand cores, it must be changed to the other end of the box for opposite hand cores. Two views of the core-box to form the recesses for links are shown between the arms of Fig. 71, and cores made in this box will be set in print D and also in a print on the segment.

The arm core-box, of which Figs. 80 and 81 are two views, should be made very strong, as it, of necessity, gets very rough usage. If this arm box is built the way I have shown, and a bolt put at each end to hold the sides together, it will stand a considerable amount of rough handling before coming apart. The depth of the box is governed by the distance of the bolt holes in the hub from the center of arm; in this case a, b, in Fig. 82, is the depth; notice also the box is made longer than it is needed for these wheels. This is done that it may be used for larger wheels. F, in Fig. 80, is made loose and is laid on top of E, which is also loose; the piece, F, is used in the box for the lower half of arm cores, and for the upper half it is taken out, because the ends of the upper arm cores must lap over the outer edge of mould in the same manner as the cope cores for the rim. (See Fig. 74).

For the band wheel, E and F are taken out and the piece in Fig. 83 substituted. The end of the box is then made to fit the inside of the rim for band wheel. The length from center of hub to C, in Fig. 73, corresponds with the distance from the center to dotted lines, d, e, in Fig. 71. The three pieces marked 1, 2, 3, are used to form the hub in arm core-box. In Fig. 80, f is a half round core print and is changed to the opposite side when piece numbered 3 is used, and on a wheel having six arms there will be twelve cores, four of each from the pieces 1, 2, 3. In order that the cope cores for the rim may fit against the sides of arm cores, two wedge pieces, g, g, are fitted against the sides of the box, making that part of the box radial, like the ends of the box in Fig. 78 are made. In making the core-boxes, a clearance, (say $\frac{1}{16}''$ on a side) should be allowed where the cores fit together at the hub and rim; nine times out of ten the moulder has to file these cores to get them in position, and it is very provoking to find, when the last arm core is being set, that it will not go in place by $\frac{1}{2}''$ or more.

The box, in Figs. 84 and 85, is to complete the outer part of the hub on each side of line a' b' in Fig. 82. h h are half round prints that will match f, in Fig. 80, when cores are set.

In Fig. 74, the dotted line, M, represents the surface of a level bed that is struck, and which is the first thing to be done towards making the mould, but before the rim can be made, the core for the end of hub must be located, the top of it being set flush with the bed, *M*. This core, which forms part of the hub, will be a guide to set the lower halves of arm cores, which is the next thing to be done. After this, the green sand part of the mould is rammed up.

The band wheel is made in much the same manner, except that in Fig. 76, the level bed is struck off at *N* and the arm cores blocked up to the proper height, and the inside of the rim rammed up.

The outside of the band wheel can be formed by cores, like *P*, being set around. When making the segment, it should be made deeper from the center rib to *N*, than it is to the cope edge. The reason for this is to allow for the piece *x*, that is made on the core, *P*, *x* being a guide for setting these outer cores, so as to give the desired thickness to rim.

ENGINE FRAMES.

HOW TO BUILD THE PATTERN TO SERVE FOR VARIOUS STROKES.

The type of frame shown in Figs. 86 and 87 is the same as used on the Corliss Engines built by the M. C. Bullock M'f'g. Co. Chicago, and is considered by mechanical engineers to be one of the best. Among a number of ways, it may be difficult to decide which is the best way to construct the pattern for both pattern maker and moulder, for in all such jobs the pattern making and moulding should be considered together.

Figs. 86, 87 and 88 are three views of this frame to be built. Fig. 87 is shown in part section towards the cylinder end of frame and Fig. 88 is a cross-section showing the two ribs on the back. The pattern is to be parted on line AB, Fig. 86, and to mould with the two ribs down. That part from AB to a, can be lifted with the cope, or it can be lifted out with an anchor plate and set in on chaplets. This pattern is also made on the same principle as the Corliss Cylinder, viz., to slide. By doing this, different stroke engines can be made with the same pattern. Provision for this change must be made between the points CD and DE, in Fig. 87, but the part between CD will only be made to slide, while the length DE can be changed by making a false end to be used for stopping off to the length required.

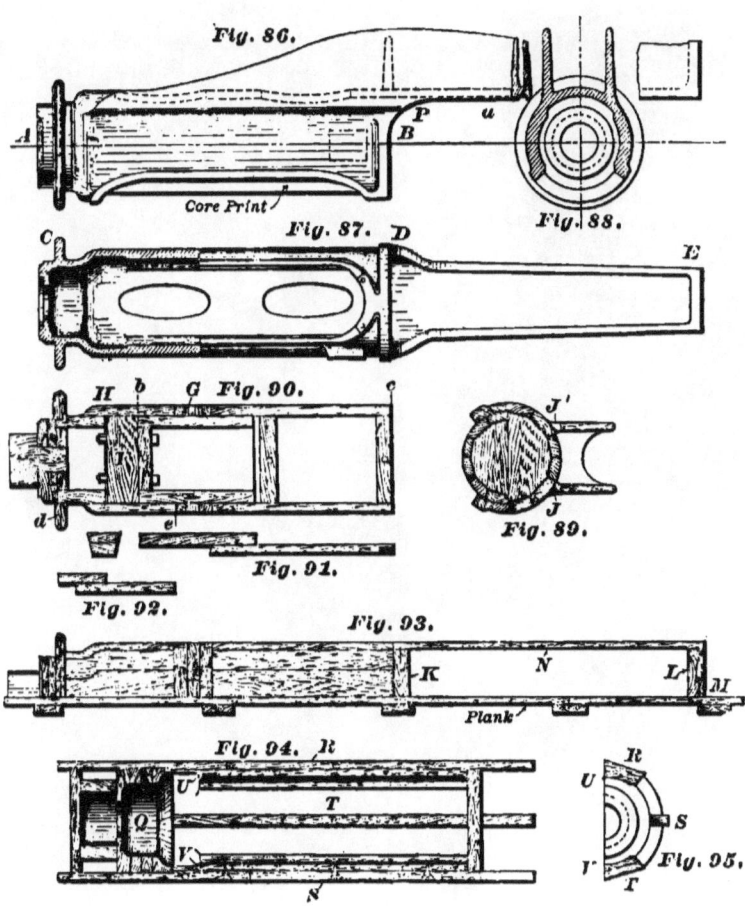

ENGINE FRAMES.

PATTERN MAKING.

We start the pattern by building up two half round pieces, one to be larger in diameter than the other; the smaller will be the diameter of the inside of frame, and will therefore serve as a core print, and the larger for the outer diameter of frame. The end view of these two parts is shown in Fig. 89, their length is from b to c, in Fig. 90. Notice that on one end of the larger part, it is turned down to the same diameter as the small part; this is for the sliding end to slide on.

Get the staves out for building the larger part, and glue on an extra thickness, forming a step as shown in Fig. 91. After building the two parts, then put them together and turn, but as one part is larger than the other, they must be balanced before turning. It will be seen that only the large piece and the end of the smaller can be turned; the remainder of the smaller piece will have to be planed off. Next get some staves out like Fig. 92 to make the sliding end H, as shown in Fig. 90. The length of this will be from d to e. The flange is glued up in two courses and fastened around this sliding end, H. The end piece that receives the cylinder should not be fastened on permanently, as one frame does for two or more sizes of cylinders, so that instead of putting on and taking off a lining, it is better to make two or or three ends for changing.

In Fig. 90 the pattern is shown arranged for the longest stroke. F is a piece bolted on the end at b to give more bearing to the sliding end, H. G is a filling piece built up in three courses so that whenever it is necessary to shorten the frame, these two pieces, F and G, can be easily taken out, and the sliding end moved up and screwed again. On referring to Fig. 86 it will be seen that the half of this sliding end that goes around the print has to be cut out circular, like the front end in Fig. 87, and moulding fitted around it. In making the moulding that is shown around the edge of frame it should be made loose from the print as far as o, o, thus making strong loose pieces, which are less likely to be broken.

That part of the frame that covers the print should now be fitted on. Having the round part of frame ready, get the ribs built on the back by first planing two flat places on the pattern where these ribs connect to the body, wide enough to take a piece that shall form the inner and outer fillets at bottom of ribs; see J, Fig. 89.

Fig. 93 shows the large part of pattern; it is laid on a plank that is surfaced; on this plank a center line is struck as a guide for getting the other part of frame in line with the body; after locating the center line on pattern to that on the plank, fix a temporary piece, K, on the end, and also a piece, L, the same height. The point, M, shows the full length of frame; these two pieces, K and L, are temporary supports for locating the straight part of frame. On these supports square up a center line from the plank. The straight part, N, should be flush with the two pieces, $J'\ J$, in Fig. 89, so that the two ribs can have a straight bearing from end to end.

Now is the best time to bore some holes through the body of pattern for screwing on the ribs. It will strengthen the pattern to glue some dowel pins through the body and down into the ribs. Having put this part together securely, proceed to fit in the bracket at P, Fig. 86, remembering that the inside of these brackets should match the inside of frame. The two lightening holes at the back, seen in Fig. 87, are cored. These cores will serve to set the main core on.

Having completed the pattern, make a half skeleton core-box for the inside of frame. This box is seen in Fig. 94, of which Fig. 95 is a cross-section. First build an end as shown at Q, and though but one-half be needed, it will pay to build a full circle and turn it. After cutting it in halves, take one and screw on three pieces, R, S, T. To R and S screw on two other pieces, U and V, for the guides. When the core is being made, the screws shown are taken out, thus leaving U and V behind to be drawn out separately. This skeleton box is arranged for the longest, and when changing length of frame all the change that the core-box will require is to fit a piece in the plain end to the desired length.

SPUR GEARS.

AND HOW THE TEETH SHOULD BE MADE.

There are so many ways of making gear patterns that it would be extraordinary to show some new way. The question is, which is the best way to make a gear pattern that will stand a reasonable amount of hammering, not be unnecessarily expensive, and so that a casting when taken from it will run smoothly. Even on a gear pattern much unnecessary time may be spent, and is spent, that does not make the pattern any better, but only more expensive. For instance, dovetailing the teeth on gears $1\frac{1}{2}''$ pitch and above, is unnecessary work. Much time can be saved by fastening the teeth on, as shown in Fig. 96. I have found this way to be cheaper and by no means inferior to dovetailing.

It is often found that after the rim has been turned and the teeth dovetailed on, that the wheel runs out; there is no doubt that the cutting of the rim and driving in the dovetails have to do with this, and yet there are men who will argue that a gear cannot be made right unless the teeth are dovetailed on.

By the method shown in Fig. 96, we have the advantage of getting a fillet at the root of the tooth, which cannot be well made on one that is dovetailed on; in this way the tooth is also strengthened and made better for moulding, which means a better casting.

In making the pattern, the blocks for the teeth should

be sawed out first, and then the segments for building the rim. These should be laid aside for awhile to allow any moisture that is in the wood to dry out, and though the stock may be considered ever so dry, it is better to do this, and the importance of it cannot be over-estimated in gear making.

In the meantime get out the arms and put them together; if the arms are of an oval section, now is the time to shape them, because it is far easier to do this before fixing them in the wheel.

Fig. 97 represents the way the arms may be put together. It will be seen that tongues are inserted in the joints where the arms come together. The grain of these tongues should not run parallel with these joints, but across.

Before building on the last two courses of segments for the rim, turn the rib that is on the inside (see Fig. 98), then build the arms in place, taking care not to fit them so tightly as to spring the rim—the remaining courses may now be laid up.

In Fig. 98 I have shown a cast iron flanged bushing; the center of arms is turned out to receive it, and is secured by one or more screws in each arm. I would suggest here that it would be a good thing to have a bushing of this kind in nearly all wheels, and that a standard sized hole be adopted. All hubs should then have a projection turned on them to fit the standard size. There would be but little trouble then in changing the hubs.

After turning the rim for the wheel, fit on the blocks for teeth, and screw them on from the inside of rim; then fit strips between, as shown; take care not to allow any glue to get between the blocks and strips when nailing and gluing on the latter, because the blocks have to be taken off again to work the teeth out.

In some cases it may be better to screw on a block and fit a strip the right width against the side of block before screwing on the next one, but this is purely a matter of choice with the pattern maker.

There is quite a variety of systems employed for laying out the profile of gear teeth, and I do not propose to enter into any discussion regarding the merits of these different systems; but much of the controversy is so hair-splitting that in practice it amounts to nothing when applied to cast gears. In some shops the teeth are laid out to "suit the eye." Of course this is entirely wrong, and, though such gears may run, there will be much jarring and friction, which means extra wear and tear, and loss of power.

For general purposes it is desirable to get that form of tooth that shall work with the least possible friction, and at the same time be interchangeable with any other of the same pitch. There are various ways of getting this. One good way is by the use of Prof. Robinson's Templet Odontograph. Fig. 96 shows how to apply the Odontograph. The full lines show it in position for marking the face of tooth, while the dotted lines show it reversed for marking the root.

To locate the Odontograph in proper position for the face of tooth, "setting" tables are supplied with the instrument, which will give the graduation on the Odontograph to set to pitch line; but this graduation is not all that is needed to set the instrument by. In Fig. 96, two dotted lines, *c* and *d*, are drawn from the *center of the tooth*, forming a right angle, *d* being radial; now, by setting the hollow edge of the Odontograph even with the line *c*, and using the graduation referred to. the location is determined for the face of tooth.

For the flank or root, two lines, *a* and *b*, are drawn from the *side of the tooth*, also forming a right angle, *b* being radial; by the aid of the line *a* and the " setting " for the flank, the Odontograph may now be set for the flank.

When a setting is found for one part of a tooth, the instrument should be screwed on a radius rod, which is moved around a center pin in the hub, so that in this Odontograph we have a ready-made templet for the teeth, and are not troubled with getting centers for the points of compasses somewhere outside the wheel or between the teeth, as sometimes happens.

While it is very desirable to make gears that are interchangeable, and which are good enough for all ordinary machinery, it must be acknowledged that the best form of tooth cannot be made by any interchangeable system; of course, I am speaking now of the epicycloidal form of tooth. For some special machinery, where it is necessary to have the best form of teeth, without regard to their being interchangeable, the Odontograph is set differently from that for the interchangeable. Prof. Robinson gives extra tables for this purpose, and herein lies the value of the Templet Odontograph, and I would recommend its being used, especially for coarse pitches.

BEVEL GEARS.

THE MANNER OF LAYING THEM OUT.

Before saying anything about the construction of these patterns, some explanation should be made about the lines that are required.

Fig. 99 is a section of the gear and pinion to be made, and gives the angle at which the two shafts are set. Fig. 100 is a face view of the gear. Bevel gears are usually made to run at right angles to each other, but when occasion requires, they can be made to run at any angle. In Fig. 99, a is the angle that is chosen; the two lines representing this angle are the first to be drawn, for on these all the others depend.

Having the angle, the next to be determined are the proportion and sizes of gear and pinion. In this case, I have made the proportion about 3 to 1, pitch 1½″. The gear is 26.8″ diameter, having 56 teeth, the pinion 9.11″ diameter, with 19 teeth. The pinion is made with an odd number of teeth, so that the teeth shall not work into the same ones of the gear at every revolution, but shall be constantly changing; the odd tooth is sometimes called a "hunting tooth."

The angle and the dimensions being settled, proceed by drawing the line b, at right angles to c. This line b is the diameter and the pitch line of the gear. From point d lay off e at right angles to f; this will be the pitch line and diameter of the pinion. Draw the center line, *g*, parallel to *f*; this locates the center, *h*, towards which all the teeth must converge. Draw *i*, *j*, *k*, the center lines of teeth, to *h*; on these lines lay off the face of the wheels, making the ends of teeth square with the center lines, *i*, *j*, *k*.

The section of the rim, arms, and hub, are now easily drawn. The teeth are laid out on larger circles than the diameters of wheels, the centers being *A* and *B*. These centers are obtained by producing the line representing the ends of teeth, or, in other words, making *m* at right angles to *j*, until it cuts the center lines of gear and pinion at *A* and *B*. The profile of the teeth must be made as if the centers of the gear and pinion were at *A* and *B*. To get the correct thickness of the tooth on the inside, the outer thickness must be laid down at *n*, running the sides toward the center, *h*. The profile of the inside of teeth is made after the same manner as the outside. Having described the method of drawing out these gears, we can pass to the construction of the patterns.

The rim is built up on a face-plate in a manner shown in Fig. 101, and when the work is thoroughly dry, the inside should be turned, getting in around at *P* as far as possible. The arms are fitted in the wheel as shown in Fig. 102, which should be done before taking the rim off the face-plate. There is a thickness of about ¾" put on each side of the arms to hold them together, also answering for fillets; see Fig. 103. Tongues should be inserted at the joints where the arms join at the center, thus making, with the pieces on either side, a strong job. It is the intention to leave the hub and ribs, *DD*, in Fig.

99, loose, so that they can be lifted with the cope, thus preventing the mould from tearing down. After fitting in the arms, the rim is chucked, as seen in Fig. 104, and finished on the outside ready for putting on the teeth. I have already described the manner in which gear teeth are put on spur gears, and these should be put on in a similar way.

Pinions may be built up hollow or solid; in this case it is not large enough to build hollow, so pieces must be glued together with the grain of the wood running with the axis of pinion, and large enough to allow the teeth to be cut out of same. For larger pinions, the pieces can be glued around the periphery, the grain of wood running the same way as the teeth, then turned off, and the teeth cut as if cutting them from the solid, as before.

HOW TO LAY OUT THE THREAD OF A WORM FOR THE PATTERN.

LET us suppose the pattern for a worm is to be made 4″ diameter, with a single right-hand thread 1¾″ pitch.

First, turn the pattern—which should be in halves—to the diameter, 4″, and to the required length, say 6″, see Fig. 106. Do not destroy the centers, because after the threads are cut, the pattern can be returned to the lathe, and the threads sand-papered on a slow speed; a much better job can be done at sand-papering this way than otherwise. After turning the pattern, wrap a piece of paper around it and cut to the exact length of circumference, and also the length of pattern. After doing this, take the paper off and lay it flat, as shown in Fig. 105.

The full lines which are laid off parallel to each other, and 1¾″ apart, represent the center line of thread, or pitch. After making these lines, take the paper and wrap it around the pattern again, and it will be found that the end marked 1 will meet the end marked 2, and 4 will meet 3, and 5 will meet 6, and so on, thus making one continuous line when wrapped on the pattern; but before gluing the paper to the pattern, mark the thickness of the thread. Be careful when you have a right-hand worm to make, that you do not make it left-hand by running the lines the wrong way; I have seen this done more than once. If it is to be a right-hand thread, the lines should run down towards the left-hand, as seen in Fig. 105. If it is to be a left-hand, they should run down towards the right.

When a double thread is required, instead of starting to draw the lines from the first division to the corner, start from the second, as shown by dotted lines. This will give a double thread, and, in fact, any number of threads can be obtained this way. If the angle of the thread is increased twice, two threads are the result; if three times, three threads, and so on until the number of threads are so many, and the angle so great, that we cease to call it a worm and begin to name it a spiral gear.

It will probably occur to some minds that if glue be used to fasten this paper to the pattern, the moisture of the glue will stretch it, so as to make the ends that meet lap over; this can be avoided and the ends need not be allowed to lap over, if a little care is exercised. Instead of spreading the glue over all the surface of the paper, put a little on the two ends that join, and also here and there on the surface; then lay the paper flat and roll the pattern carefully on it, the pattern gathering up the paper around it as it is rolled. If a spiral gear is to be laid out this way—and it can be—the two ends of the paper should meet exactly.

WORM WHEELS.

THE WAY TO GET THE ANGLE OF TEETH, AND THE MANNER OF FASTENING THEM ON.

A WORM-WHEEL pattern is not an easy thing to make by any means; it is that kind of a job on which a good man is apt to go astray.

Figs. 107 and 108 represent a wheel to be made; the worm is shown in gear, and as I have already referred to it, I will confine my remarks to the wheel pattern, except that reference will have to be made to the angle of thread of worm.

The dimensions are the first thing to be determined. Let the wheel be 21¾" diameter, 1¾" pitch, 39 teeth, form of tooth, involute, laid out with the aid of Prof. Willis' Odontograph.

Pattern makers generally shape the thread of a worm pattern the same as the tooth of the gear in which it is to work; but the sides of the thread should be straight, at an angle of 75 degrees with the axis of the worm. This is correct for an involute, because that is what a rack tooth would be, and the worm is similar to the rack in this case. A section of the wheel pattern is shown in Fig. 109; the rim is first built up and turned straight on the outside, the pattern being parted on line *AB*.

A little thought will have to be exercised in building and turning the rim so as not to do any unnecessary chucking. In the first place, the face plate for turning it should be about the same diameter as the rim, so as to allow the ends of the teeth to be turned off. Lay each half up separately, with the parting joint of pattern against the face plate, rough off the outside and finish the inside of both halves, care being taken to turn the inside of both halves the same diameter, because of the chucking. This done, the rings are chucked by the inside, and, to do this, segments are nailed on the face plate. The outside of the rim should now be finished to the size required, and blocks for the teeth fitted on the periphery. These blocks should be fitted and glued on with the grain of the wood running at the same angle that the teeth are to be at the pitch line, the width of each block being the same as the pitch. The way to get the angle is shown in Fig. 110, y representing the angle of tooth at the pitch line. These blocks are fastened on before taking the first half out of the lathe, and a groove turned on the joint for locating the halves concentric with one another. When the second half is chucked and turned and the blocks for teeth fixed on, as in the case of the first half, a projection is turned on the joint of it to fit the groove on the joint of the other. The two halves should now be put together, and the blocks for the teeth turned off on the ends, and finished, as seen in section Fig. 109. A good surface for marking out the teeth may be made by varnishing these blocks with yellow varnish.

In Fig. 107, a part of the pattern is shown cut through line AB, Fig. 109; the other part, with the three teeth represents the outside and ends of the teeth. To be theoretically correct the teeth should be thinner on the ends than on the parting line, AB, but in practice, this is not generally considered, because it would make it difficult to draw the pattern if made that way; the teeth in cast gears are therefore made the same thickness on the line C as on the pitch line D in Fig. 107.

The Odontograph for locating a center for marking out the teeth is shown at E; the instrument is nothing more than an angle of 75° divided off in $\frac{1}{4}''$ spaces on one side for $4''$ in length, the $\frac{1}{4}''$ spaces being subdivided. Zero on the instrument is placed at a point on the pitch line, and the plain side is set to a radial line; the radius of the gear is then read off on the graduated side, which will locate a center from which to strike the tooth. In this case, the wheel being $21\frac{3}{4}''$ diameter, the radius is $10\frac{7}{8}''$; read off 11, which is near enough for all practical purposes.

To lay out the teeth on the pattern at the proper angle, space off half the pitch from one of the joints of the blocks, on the pitch line each side of the wheel. If these pieces for the teeth have been made to the angle of y, Fig. 110, these points on each side will be the starting points for dividing the teeth, and will give the angle of tooth.

If the teeth are marked out also on the joint of one-half of the pattern, and trimmed down to the line, it will be an additional guide to cut the teeth by.

The fact of the arms of the pattern being in halves makes them thin, and therefore weak; but if the joints in the center be tongued, and the hub on each side glued and screwed to the arms, it will strengthen them considerably.

SWEEPING STRAIGHT WINDING DRUMS.

Fig. 111 shows one way of sweeping up hoisting drums. The usual way is to allow one or two fingers for sweeping the groove to travel the whole length of drum by means of a nut running on a long screw. It will be seen that the long screw is discarded and that the sweep is made the whole length that the drum is to be. This sweep is required to travel up and down only one thread, which will make a continuous thread, the same as if two fingers traveled the whole length by means of the screw. The set screws, E and F, are free from the spindle, so that the sweep may rise and fall as it is pulled around. It rises by means of a pinplate, B, bolted to the bottom of the sweep, and which works in a spiral groove cut in the hub, A; this groove must be cut the same pitch that the groove of drum is intended to be.

Figs. 112 and 113 are other views of A and B. The spindle is kept from turning by the set screw, H, being set against it; at the same time, the projection on the hub, A, fits into the slot G, in Fig. 112.

After sweeping the grooved part of drum, the part C of the sweep may be taken off and a piece, like D, fastened to the upper cross-piece; this is for striking the flange of drum and a step around the top for a guide by which to set the cope. While sweeping the groove the spindle remains stationary, but when sweeping this

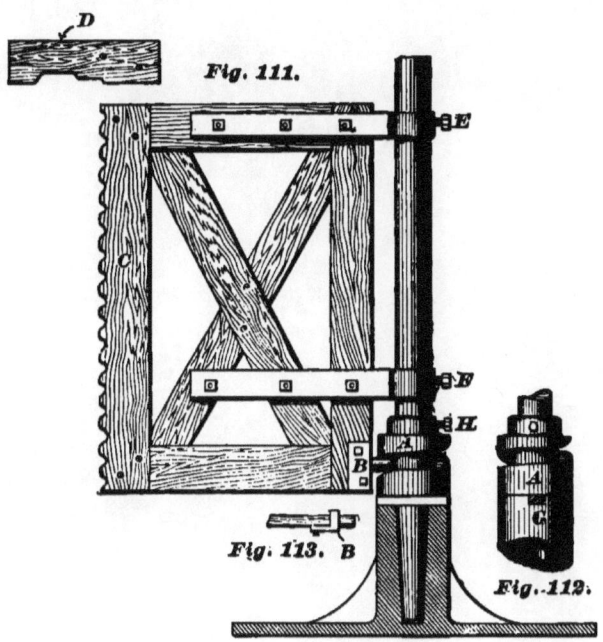

SWEEPING DRUMS.

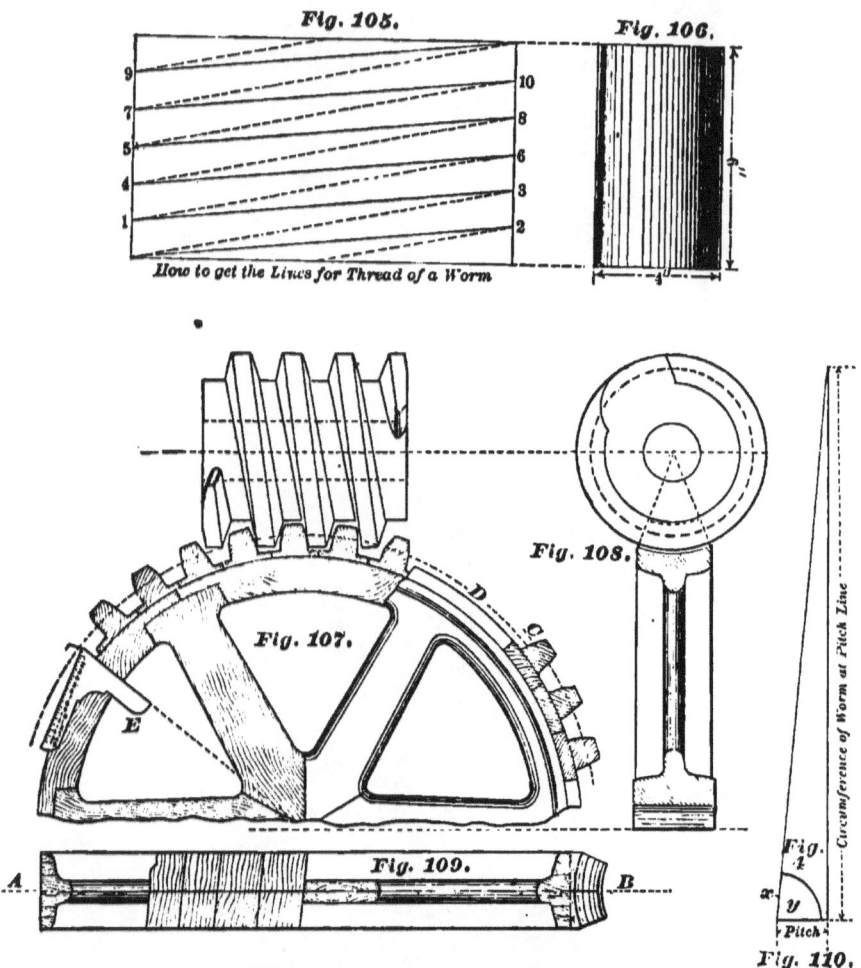

How to get the Lines for Thread of a Worm

WORM WHEEL PATTERN.

flange and step, it must revolve with the sweep; to do this, the set screws, E and F, must be tightened, the pin-plate, B, taken off the sweep, and the set screw, H, loosened; this will allow the spindle to revolve in its taper socket in the ordinary way.

For drums over 6 ft. in diameter, it will be found necessary to support the top of the spindle with a temporary cross-beam, and also, to have a counterbalance weight to help the rising of the sweep as it is pulled around.

MAKING WINDING DRUMS FROM PATTERNS.

METHOD OF CUTTING THE GROOVE.

FIG. 114 shows a section of a Grooved Winding Drum about 3 ft. diameter, and Fig. 115 shows the way pattern may be constructed.

D is a 4x4" stick; on it three discs, A, B, C, are fastened and screwed by brackets. Lagging to form the shell is screwed around these discs, while the spider in the end is loose and just laid on disc A, so that it can be lifted with the cope. C should be made about $\frac{3}{4}''$ smaller in diameter than A, to allow the core, shown in Fig. 114, to be easily lifted; this $\frac{3}{4}''$ taper on the drum will not be noticeable when cast, neither does it affect anything.

Fig. 116 represents a piece of the lagging that is screwed on the discs, A, B, C. Fig. 117 is given to show how to get the proper angle of the groove; the distance, x, is the circumference, a the pitch, and y the angle that the grooves are to be cut. Fasten two pieces together in the form of a **T** square, like Fig. 118, for marking the lines on Fig. 116, and let the blade be circular, to fit the face of the lagging. Also make a templet the same length and thickness of the drum, to mark off the grooves on the edge of lagging.

As the outside is being rammed up, the screws that hold the lagging must be taken out, so as to allow A, B,

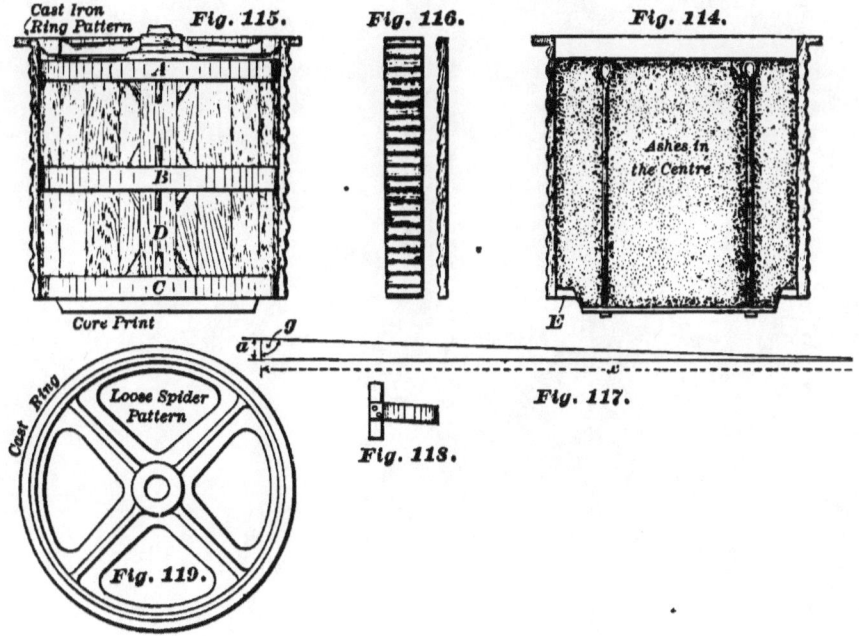

PATTERN FOR WINDING DRUM.

C, and *D* to be removed after the cope is lifted. Previous to ramming up the core, a flange, *E*, Fig. 114, is placed in the bottom of the mould; this flange is made in segments and is for bolting the brake wheel to the drum. After the inside is rammed up and lifted out, this flange, *E*, and the lagging are drawn in and the groove finished off and the skin dried.

Some may say that a pattern such as I have described is very costly; in reply I would say that it would soon pay for itself by the difference in the cost of sweeping and moulding, but drums of larger diameter should be swept up, as they generally are.

MAKING SHEAVES FROM CORE-BOXES.

I PROPOSE to give under this head some ideas for making sheaves of various kinds, and will first give the way for making large ones from core-boxes.

The style of sheaves, shown with cast arms, is that which is used to transmit power on cable car roads. The groove is made to receive segments that are bolted on, and of which more will be said further on. The right hand side of Fig. 120 shows one arm and a portion of the rim; as will be seen, this sheave is in halves, and arranged for bolting together; joint E is to be finished so that stock for planing must be allowed at E. In Fig. 121 two sections are shown; that on the right is a sectional view of the mould through the arm-core, AB, and that on the left through CD. These two sections represent the mould closed, but in Fig. 120 I have shown the mould open, with the arms, F and G, set. The lower part of rim is made in green sand, a segment pattern being used to form it. This segment pattern is shown to the left in Fig. 120, H being a section of it. Two battens, a, b, are screwed on, which run to the center, the ends being cut to fit around the spindle.

The lugs at the joint, for bolting the rim together, must be right and left hand on the segment pattern, using one when starting and the other when finishing. These lugs are screwed on temporarily. Care must be

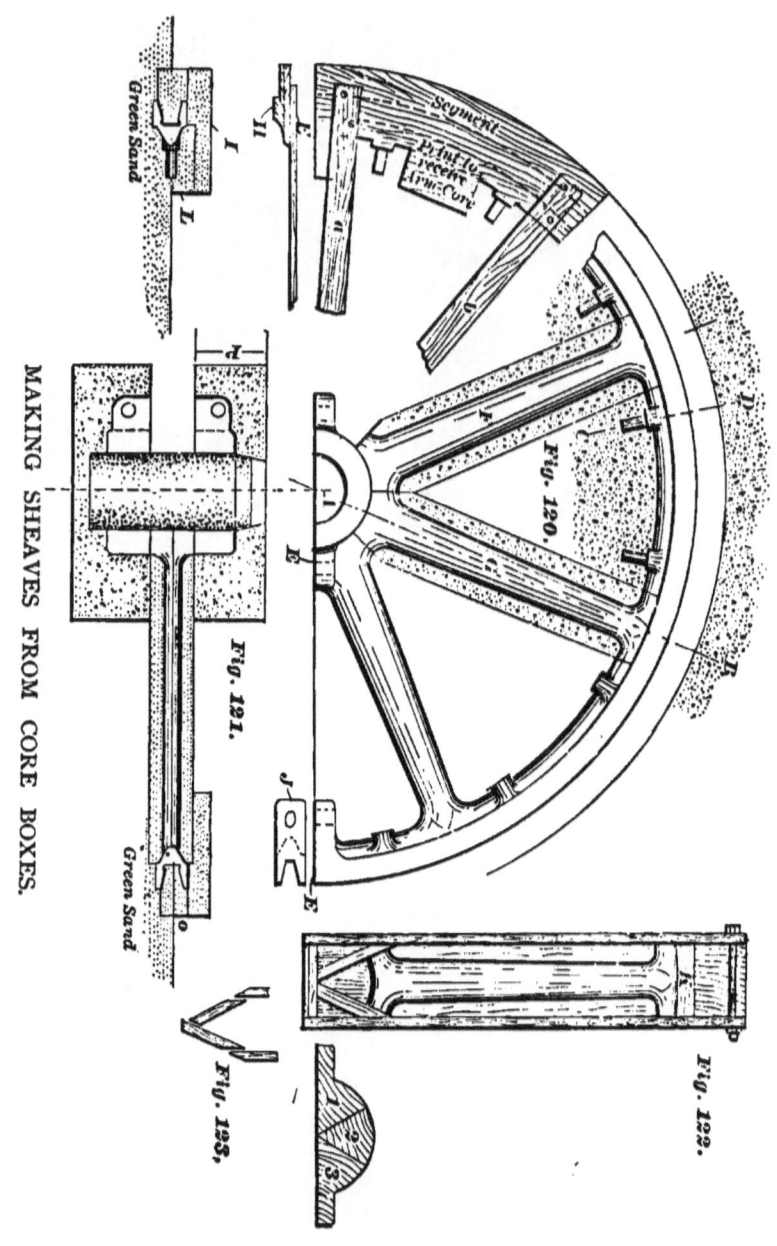

MAKING SHEAVES FROM CORE BOXES.

MAKING SHEAVES FROM CORE BOXES.

Fig. 126.

Fig. 124.   Fig. 125.

taken to divide the arms off accurately on a level bed. After marking the center line on the bed for each arm, move the segment around on the spindle and apply the center line of the core print for the arm core to the center lines on the bed, and, as lines on sand soon disappear, it will be well to drive a small stake on each side of the print for arm core, so that, when moving the segment to each division, the arm print can be set down between the stakes, and thus insure accuracy. It is particularly necessary to divide these arms off equally, so that the bolt-holes shall match those in the segments that go in the groove, for it is the intention that all the holes, both for bolting together the sheave and segments, shall be cast in, and to compensate for any difference that may occur, the holes are made a little long in the rim and lugs, as seen at J, Fig. 120.

Fig. 122 is the arm core-box, and I will again remind the pattern maker that it should be made strong, or it will come apart, as I have often seen, and then there is trouble about the cores not coming together at the center as they should.

Fig. 123 shows the way some pattern makers construct these large arm core-boxes; the result is, they are rammed apart, as shown, and then the poor core maker is accused of using the core-box roughly. Just as in the case of arm box for fly-wheels, so this box should be made a little longer than is necessary for the present job, so that the end, K, can be moved out and the arm lengthened for a larger wheel whenever it may be needed. The interchangeable pieces, 1, 2, 3, form the hub.

Fig. 124 is the core-box from which the cope cores, L,

in Fig. 121, are made. The same lugs, *J*, that are used on the segment pattern, can be used in this box for two cores, using one right hand and one left hand. The box is shown arranged for the first core on the left. *M* is a loose piece half the width of the core print that receives the arm core. It will be seen that this box is made longer than the core is needed. This is to enable us to change ends with the loose piece, *M*, when making a core the opposite hand. The length of this box is from the center of the arm to the joint *E*, but the $\frac{1}{4}''$ stock, which is allowed for planing the joint at *E*, makes the box $\frac{1}{4}''$ longer than the eighth part of the half sheave, and therefore $\frac{1}{4}''$ too long for the other cores, so that a $\frac{1}{4}''$ piece must be put in the end of the box after making two cores with the lug, *J*.

Fig. 125 is the box for the groove cores. The section of this box shows it arranged for making the cores in two parts, to be pasted together at *o o*, Fig. 121. The cope part of this core is a little different from the bottom part; a loose piece, *g*, in Fig. 125, is fitted in the bottom of the box, to be left in for the bottom part of core and taken out for the cope part. The groove that this core is to form is turned on the two sides, but not in the bottom; tool clearance should therefore be allowed on each side in the bottom to accommodate the turning. This is seen in the section at *N*.

I have not shown any core-box for the slab core, *I*, as it is nothing but a plain core, and the box is simple and needs no explanation as to making it.

Fig. 126 is the core-box for the hub. It is made the depth of *P*, in Fig. 121. This view gives all the explan-

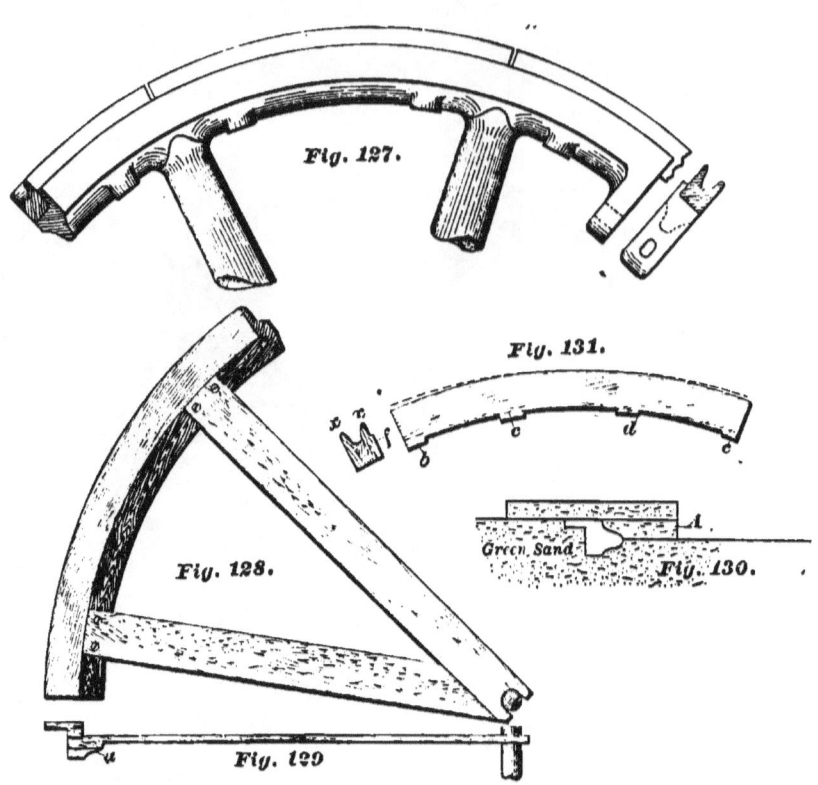

MAKING SHEAVES FROM CORE BOXES.

ation that is necessary as to the way to make it. The faces, *E*, of the hub and lugs are covered with slab cores.

Another type of sheaves is shown in Fig. 127. This style can be made considerably cheaper than those I have already described. Dispensing with the groove that receives the segments makes it very simple to mould, and also easier for turning the periphery of casting. The plan for forming the arms and the hub is the same here as in Fig. 120. The segment pattern for making the green sand part of mould is seen in Fig. 128, of which Fig. 129 is an end view. The pieces that run to the center are not screwed on top of segment in the usual way, but on the step that is made in the segment at *a*, Fig. 129. By making it in this way, a large part of the rim can be made in green sand, as shown in Fig. 130.

The core-box for core *A*, is made in very much the same way as in Fig. 124. The segment that is bolted on the periphery of this sheave is moulded edgewise; there are chipping strips on the inside at *b, c, d, e*, Fig. 131; on the side, *f*, stock for planing is allowed, so that the segment may set straight against the flange of the wheel.

I once had a little experience with some of these segments. When the first lot of those in Fig. 131 were being fitted on the wheel, it was found they had straightened somewhat, just as represented by the dotted lines; it was evident that these segments straightened in cooling, the two thin flanges, $x\ x$, cooling first and pulling the casting out of its true circle; in the next, I took care to allow for this when making the pattern.

When there is a number of these sheaves to make, instead of closing the top with slab cores, as I have shown, it would pay to make a cast iron half ring with which to cover the top. This half ring should have a number of spikes on one side, and on it a thick coat of loam, struck off level, dried, and blacked.

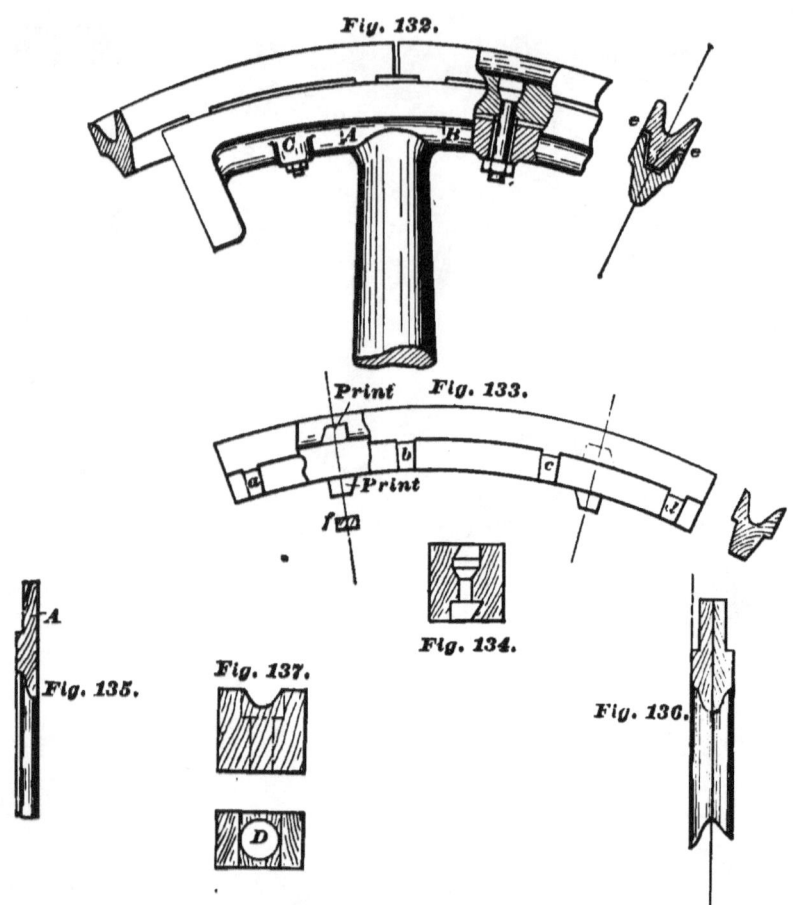

MAKING SHEAVES FROM PATTERNS.

MAKING SHEAVES FROM PATTERNS.

THERE is not much scheming required to make a pattern for a sheave, such as shown in Fig. 132, and yet, to show the way it should be made, may not be entirely out of place here, as I want to bring in a few points that have not hitherto been considered.

I have said that into the groove of this style of sheave are bolted segments that take the cable. The advantage of this arrangement is evident, as it allows the segments to be renewed when worn out. I have shown in Fig. 132 a part of the rim and a cross-section of the sheave; this shows the manner of bolting the segments to the sheave. The groove, into which the segments are bolted, is to be turned, but the groove of the segment is left rough.

Chipping pieces are cast on each side of the segment, as seen at a, b, c, d, Fig. 133, because it is intended that the segment shall not bear in the bottom of the groove, but only on the chipping pieces by the sides, and at $e\ e$; see cross-section, Fig. 132.

Fig. 133 shows the pattern of the segment, and is made to be moulded on the edge, the groove being in the cope; it is desired to cast the bolt-holes, and care must be taken in spacing them off, because they are wanted to match those in the sheave, which are also cast in. I have marked the core prints for these holes; the bottom print is made something like the cope print—

oblong—as shown at f; this is done in order that the core may stand in the mould more securely while the cope is being closed. If a round print were used just the size of holes, the cores would be top-heavy and difficult to locate in the mould, hence the necessity of making the bottom print as shown. The core-box for this bolt-hole should be made as shown in Fig. 134.

It is understood that these sheaves are bolted together in halves, so that in making a pattern, only one-half will be required. Proceed by building up and turning a whole ring, of which Fig. 135 shall be the section; A is the print for carrying the groove cores. After turning the ring, cut a stick the exact length of the inside diameter—this will be a gauge to see whether the ring has sprung after being cut in two, and, if it has, to bring it back to the gauge when fastening in the arms. After sawing the ring in two, glue and screw it together strongly, as seen in Fig. 136; but before doing so, it must be remembered, as before, that stock for planing must be allowed at the joints where the ring is bolted together, so that the pattern shall be $\frac{1}{4}''$ over the half circle. In order that this may be, the ring should not be cut exactly in halves, but $\frac{1}{4}''$ one side of the center, making one part about $\frac{1}{2}''$ short, not reckoning anything for saw cut—with saw cut would probably be $\frac{5}{8}''$ short. Now, if after sawing the ring, two of the ends be brought together, it will only be necessary to build on one end of one of the sections. Having done this, the ring is ready to have the arms fitted in, which should be done by letting them in the rim, as represented by dotted lines at $A B$, Fig. 132.

Care must be taken not to fit the arms in so tightly as to spring the ring out of round; this can be very easily done. After locating the arms, bore two $\frac{3}{4}''$ holes from the outside of ring into each arm, and glue in hard wood dowel pins; this will make a strong job. The small bosses, of which one is shown at *C*, Fig. 132, are turned and sawed out with a narrow band-saw to fit over the rim. This is done by inserting each boss in a block with a hole through it the size of the boss; two views of this block are seen in Fig. 137. The boss is fixed in the hole *D*, and sawed to the shape of the inner part of rim. Of course, the block is fitted over the rim first, to act as a guide for sawing them out.

The core-box for the groove need not be made with loose piece in the bottom, as in case of forming these sheaves with cores, because the cope closes down on top of print, and not on the dotted line, Fig. 136.

SHEAVES WITH WROUGHT IRON ARMS.

AN ORIGINAL WAY OF MAKING THE HUB.

THE style of sheave shown in Fig. 138 is used extensively in mines for carrying rope; the arms, which spread on either side, act like stay-rods to the rim, making it very rigid sidewise, at the same time forming altogether, a light, but strong, sheave.

To the left of Fig. 138 is shown a section of the rim with wrought iron arms cast into it; to the right, a section of the cores which form the rim; and at the center, a section of the cores forming the hub.

The lower part of the mould is formed with green sand, the segment shown in Fig. 139 being swung around from the center, C. The cope is formed with cores made from box shown in Fig. 140. Fig. 141 is a cross-section of this box.

While this is a good way to make sheaves of large diameters, for those under 8 ft. diameter a full ring is probably a better way, providing the ring can be stored so as to lie flat on its side, instead of standing on its edge; for, having no arms, a large ring standing edgewise would soon become oval.

The cores forming the groove are made in halves from the box, of which Figs. 142 and 143 are two views; these cores are pasted together at the joint, A, Fig. 138. Four round cores are made to form the hub; these are

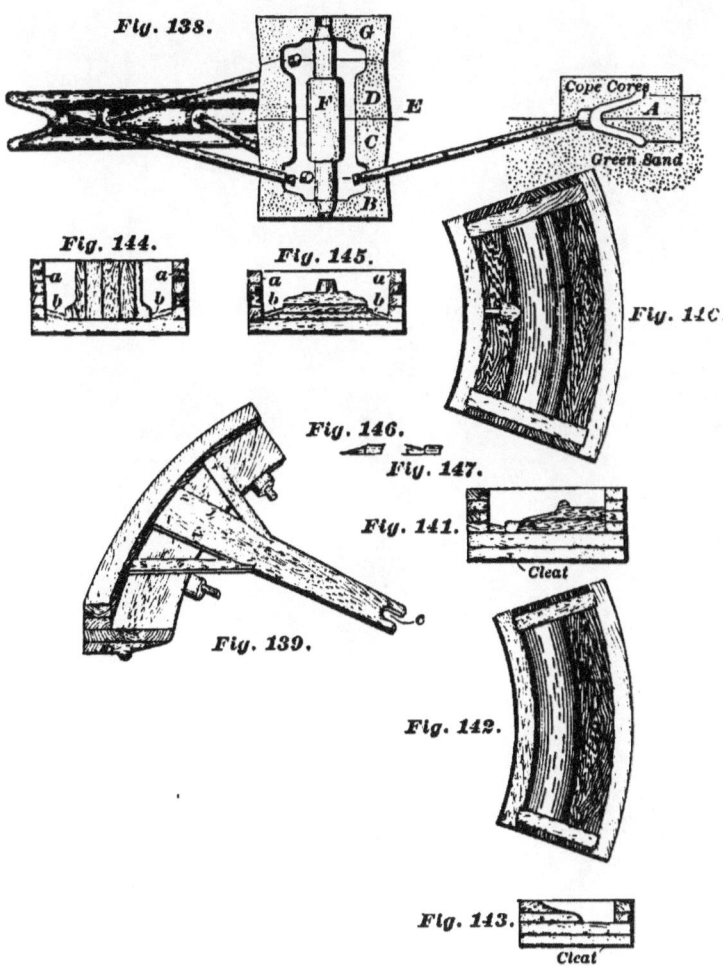

SHEAVES WITH WROUGHT IRON ARMS.

set one on top of the other after making the lower part of the mound with the segment. The lower hub core, B, through which there is a hole, should be set over the pin from which the segment has been swung around. This will locate the hub concentric with the rim. Half of the arms should now be set in the mould, after which, the two middle cores, C and D, are located. When C and D are being pasted together at the joint, E, care should be taken to get the holes that receive the arms exactly midway between those in the lower part. The center core, F, should now be set, then the balance of the arms and the top core, G. When pouring these sheaves, the rim is allowed to shrink all it will before pouring the hub, and in large ones, the hub is not poured until the following day.

Figs. 144 and 145 are sections of the round core-boxes for the hub cores, CD and DG. Plenty of draught should be made on the inside of these two boxes at $a\ a$; the half round prints for the arms are shown at $b\ b$.

The small bosses shown in Figs. 146 and 147 are used in cope core-box, Fig. 140; the prints on these small arm bosses vary; the cores which cover those arms running upward from the hub, should have Fig. 146 in the box, and those which run downward, Fig. 147. The bosses on the segment are made similar.

A MACHINE

FOR

SWEEPING CONICAL DRUMS.

DESIGNED BY THE AUTHOR.

It may not be understood by some why a winding drum is sometimes made conical instead of a straight cylindrical form, and it may not be entirely out of place here to explain the reason, for the benefit of such.

Conical drums are used for winding heavy loads from deep mines. When the skip or load is at the bottom of the mine, ready to be hauled up, the winding on the drum begins at the small end, and, as the rope does not wind as fast on the small end as it does on the large end of the drum, it allows the slack rope in the shaft to be gradually taken up at the starting, and also prevents the load from starting too suddenly. The engines also gain a decided advantage when winding with conical drums, because, instead of the winding being started at full speed, it gradually increases, thus giving the engines a better chance to do their work.

It is scarcely necessary to inform my readers that it requires a great deal more skill to build and properly secure a mould for a large conical drum than it does to mould a grate bar; but there is a class that stands so high in the engineering profession, that to them all foun-

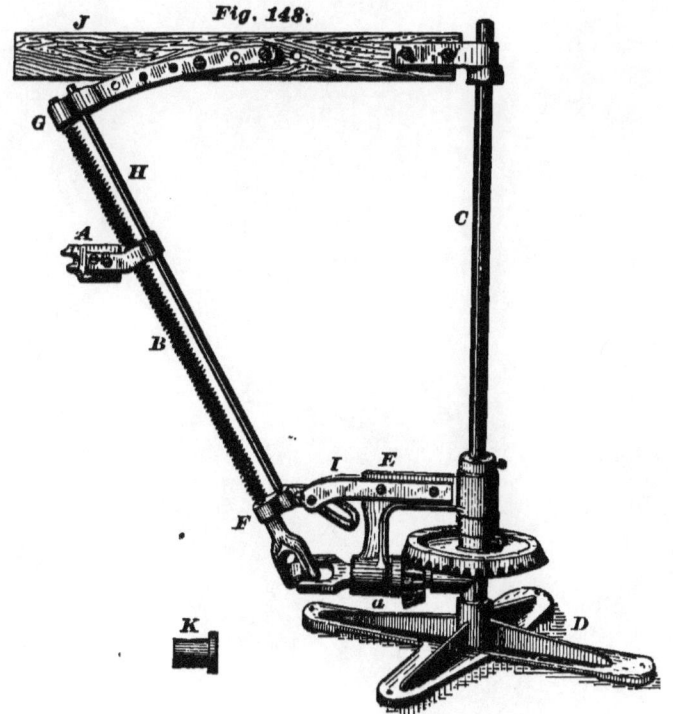

SWEEPING CONICAL DRUMS.

dry work is just a little above unskilled labor—something requiring more brute force than anything else. Such ideas, though, do not prevail among our genuine and practical engineers; they are only found among the cleverly ignorant.

All those who have much to do with the machinery business know what an amount of consultation and scheming is necessary before some jobs in a foundry can be started, and then how it requires men of good sound judgment to execute the work.

The building of a mould for a large conical drum is one of these jobs. The way of sweeping the groove, an arrangement for which I propose to describe, is only a small item of the work.

In Fig. 148, A is the sweep that travels up and down the screw, B, as it is pulled around. The spindle, C, is secured to the cross, D, at the bottom; the bevel gear is fast on the spindle, two set-screws in the hub holding it in place; the bracket, E, is loose, and turns on the spindle; it has a bearing at a, in which the pinion shaft runs; the end of this shaft is carried by a tee piece that turns on the spindle. The pinion shaft and the screw are connected by a universal joint, while the screw is carried by two adjustable curved pieces, F and G. Guide-rod, H, keeps the nut from turning on screw, B; arm, I, fits over the bracket, E, and carries the curved piece, F; this arm is also adjustable.

Now, it will be clearly seen that, if the arm, J, and the bracket, E, are pulled around, it will cause the pinion and the screw, B, to turn, thus making the sweep, A, to travel a certain distance every time it goes around. The gears determine the pitch of the groove to be swept; if the proportion of the gears are three to one, and the screw $\frac{1}{2}''$ pitch, then the sweep will travel $1\frac{1}{2}''$ at every turn, making a groove $1\frac{1}{2}''$ pitch. When any other pitch is required, the gears must be changed for those of a different proportion, for instance, for a $2''$ pitch drum the proportion of the gears would be 4 to 1. Bracket, E, is made so as to permit the use of gears of different sizes.

When a drum is wanted with a left-hand groove, the gear on the spindle is turned upside down, and located under the pinion instead of over it. The machine is also arranged so that a drum of any angle can be swept. This is done by loosening the bolt that holds part F in place, and by taking out those in the upper part, G, thus allowing the screw to be swung at any angle from the center of the universal joint.

The reason for making the arm, I, separate from the bracket, E, is obvious; it is to give a better chance for adjusting the lower part of the machine than the swinging of the screw gives. The pinion shaft runs in close to the upright spindle, so that when the set-screw in the pinion is loosened, the shaft can be pulled out to the required distance, and the set-screw in pinion be tightened again. When this is done, it will be found necessary to bolt on the flanged sleeve, K, to the end of bracket, E, between the universal joint and the bearing, a. For building and sweeping up the mould roughly, the screw and pinion shaft can be disconnected entirely and a plain sweep made, bolting it to the upper and lower arms, J and I.

The engraving only represents the model which I

made of this machine; the details of the machine proper will vary somewhat. For instance, where there are solid boxes on the bracket, *E*, for the spindle and pinion shaft, there should be caps, so as to make it easier to disconnect the parts. The universal joint should also be made separate from the screw and pinion shaft; many other items would need changing when building a machine to do the work.

GEAR TEETH.

In the following pages there are a number of teeth laid out, full size, from one inch pitch to three inch, advancing by quarter-inches.

There are fourteen separate teeth in each pitch, suitable for gears having from fourteen to eight hundred teeth; they have been laid out from Prof. Robinson's Templet Odontograph and are interchangeable. The clearance allowed between the teeth is $\frac{1}{20}$ of the pitch, or in other words, the space is $\frac{55}{100}$ and the thickness of tooth $\frac{45}{100}$ of the pitch; the height of the tooth is $\frac{7}{10}$ of the pitch, and the distance from pitch line to top $\frac{30}{100}$ of the pitch. This is the proportion used for general purposes.

A templet can be made from any of these teeth and fastened on a rod and used in the same way as the Odontograph is in Fig. 96. It would be impossible in a book of this kind to give the profiles of gear teeth which would serve all cases, so that I have confined myself to the system that is generally adopted and known as the Interchangeable System, that is, all spur gears of the same pitch made under this system will run together. For special gearing and bevel gears other settings are preferred; those which I have taken are on each tooth, the setting for the flank being marked on the inside of pitch line and that for the face on the outside. The thickness of each tooth at the bottom and top, and also at the pitch line, is correct, so that by the aid of the settings marked, the odontograph can be easily applied for striking the curves on a piece of sheet zinc, from which a templet tooth is usually made.

The numbers show how many different size wheels can be made with same size tooth; for instance—42 to 47 means that the same shape tooth will answer for gears which are to have from 42 to 47 teeth.

On suceeding pages, at the end of the book, will be found plates, in which some of the teeth are shown in gear, together with the way they should be made.

Table of the Diameter of Wheels at the Pitch Circle, from 11 to 300 Teeth.

Number of Teeth.	Pitch of the Teeth.									
	inch. 1¾.	inch. 1⅞.	inches. 2	inches. 2⅛.	inches. 2¼.	inches. 2 ½.	inches. 2¾.	inches. 3.	inches. 3¼.	inches. 3½.
11	0 6¼	0 6⅝	0 7	0 7½	0 7⅞	0 8¾	0 9¾	0 10⅝	0 11⅜	1 0¼
12	0 6¾	0 7⅛	0 7⅝	0 8⅛	0 8⅝	0 9⅝	0 10⅝	0 11½	1 0⅜	1 1⅜
13	0 7⅜	0 7⅞	0 8⅜	0 8⅞	0 9⅜	0 10⅜	0 11½	1 0½	1 1⅝	1 2½
14	0 7⅞	0 8½	0 9	0 9½	0 10	0 11¼	1 0⅜	1 1½	1 2½	1 3⅝
15	0 8½	0 9	0 9⅝	0 10¼	0 10¾	1 0	1 1¼	1 2⅜	1 3½	1 4¾
16	0 9	0 9⅝	0 10¼	0 10⅞	0 11½	1 0¾	1 2	1 3⅜	1 4½	1 5⅞
17	0 9⅝	0 10¼	0 10⅞	0 11½	1 0¼	1 1½	1 2⅞	1 4¼	1 5½	1 6⅞
18	0 10	0 10¾	0 11½	1 0¼	1 0⅞	1 2⅜	1 3¾	1 5¼	1 6⅝	1 8
19	0 10⅝	0 11⅜	1 0⅛	1 0⅞	1 1⅝	1 3⅛	1 4⅝	1 6¼	1 7⅝	1 9⅛
20	0 11¼	1 0	1 0¾	1 1½	1 2⅜	1 4	1 5½	1 7⅛	1 8¾	1 10¼
21	0 11¾	1 0½	1 1½	1 2¼	1 3	1 4¾	1 6⅜	1 8⅛	1 9¾	1 11⅜
22	1 0⅜	1 1⅛	1 2	1 2⅞	1 3¾	1 5½	1 7¼	1 9	1 10¾	2 0½
23	1 0⅞	1 1¾	1 2⅝	1 3½	1 4½	1 6⅜	1 8	1 10	1 11¾	2 1⅝
24	1 1½	1 2⅜	1 3⅜	1 4¼	1 5¼	1 7⅛	1 9	1 10⅞	2 0⅞	2 2¾
25	1 2	1 2⅞	1 3⅞	1 4⅞	1 6	1 8	1 9⅞	1 11⅞	2 1⅞	2 3⅞
26	1 2½	1 3½	1 4½	1 5½	1 6⅝	1 8¾	1 10¾	2 0⅞	2 2⅞	2 4⅞
27	1 3	1 4⅛	1 5¼	1 6¼	1 7⅜	1 9½	1 11⅝	2 1¾	2 3⅞	2 6⅛
28	1 3⅝	1 4⅝	1 5¾	1 6⅞	1 8	1 10¼	2 0½	2 2¾	2 4⅞	2 7¼
29	1 4⅛	1 5⅜	1 6½	1 7⅝	1 8¾	1 11⅛	2 1⅜	2 3¾	2 6	2 8⅜
30	1 4½	1 6	1 7⅛	1 8¼	1 9½	2 0	2 2¼	2 4⅝	2 7	2 9⅜
31	1 5⅜	1 6½	1 7¾	1 9	1 10¼	2 0¾	2 3⅛	2 5⅝	2 8	2 10½
32	1 5⅞	1 7⅛	1 8⅜	1 9⅝	1 11	2 1½	2 4	2 6½	2 9⅛	2 11⅝
33	1 6½	1 7¾	1 9	1 10⅜	1 11⅝	2 2¼	2 4⅞	2 7½	2 10⅛	3 0¾
34	1 7	1 8⅜	1 9⅝	1 11	2 0⅜	2 3	2 5¾	2 8½	2 11⅛	3 1⅞
35	1 7½	1 9	1 10⅛	1 11¾	2 1	2 3⅞	2 6⅝	2 9½	3 0¼	3 3
36	1 8	1 9½	1 10⅞	2 0⅜	2 2	2 4⅝	2 7½	2 10⅜	3 1¼	3 4⅛
37	1 8⅝	1 10	1 11½	2 1	2 2½	2 5½	2 8⅜	2 11⅜	3 2¼	3 5¼
38	1 9¼	1 10¾	2 0¼	2 1¾	2 3¼	2 6¼	2 9¼	3 0¼	3 3⅜	3 6⅜
39	1 9¾	1 11⅜	2 0⅞	2 2⅜	2 4	2 7	2 10⅛	3 1¼	3 4⅜	3 7⅝
40	1 10⅜	1 11⅞	2 1½	2 3	2 4⅝	2 7⅞	2 10¾	3 2¼	3 5⅜	3 8½
41	1 10⅞	2 0½	2 2⅛	2 3¾	2 5⅜	2 8⅝	2 11⅞	3 3⅛	3 6⅜	3 9⅝
42	1 11½	2 1	2 2¾	2 4½	2 6	2 9⅜	3 0¾	3 4⅛	3 7⅜	3 10¾
43	2 0	2 1⅝	2 3⅜	2 5	2 6¾	2 10¼	3 1⅝	3 5	3 8⅜	4 0
44	2 0½	2 2¼	2 4	2 5¾	2 7½	2 11	3 2½	3 6	3 9½	4 1
45	2 1	2 2⅞	2 4⅝	2 6½	2 8¼	2 11¾	3 3⅜	3 7	3 10½	4 2¼
46	2 1⅝	2 3½	2 5¼	2 7⅛	2 9	3 0⅝	3 4¼	3 7⅞	3 11½	4 3¼
47	2 1⅞	2 4	2 6	2 7⅞	2 9⅝	3 1½	3 5⅛	3 8⅞	4 0⅝	4 4⅜
48	2 2¾	2 4⅝	2 6½	2 8½	2 10⅜	3 2¼	3 6	3 9⅞	4 1⅝	4 5½
49	2 3⅜	2 5¼	2 7¼	2 9⅛	2 11	3 3	3 6⅞	3 10⅞	4 2⅝	4 6½
50	2 3⅞	2 5⅞	2 7⅞	2 9¾	2 11¾	3 3¾	3 7¾	3 11¾	4 3¾	4 7¾
51	2 4½	2 6½	2 8½	2 10½	3 0½	3 4½	3 8⅝	4 0¾	4 4¾	4 8⅞
52	2 4⅞	2 7⅛	2 9⅛	2 11⅛	3 1¼	3 5⅜	3 9½	4 1⅝	4 5¾	4 10
53	2 5⅜	2 7⅝	2 9¾	2 11⅞	3 2	3 6¼	3 10½	4 2⅝	4 6⅞	4 11
54	2 6	2 8¼	2 10⅜	3 0½	3 2⅝	3 7	3 11	4 3½	4 7⅞	5 0⅛
55	2 6⅝	2 8⅞	2 11	3 1¼	3 3½	3 7¾	4 0⅛	4 4½	4 8⅞	5 1¼
56	2 7⅛	2 9⅜	2 11⅝	3 1⅞	3 4⅛	3 8¼	4 1	4 5½	4 9⅞	5 2⅜
57	2 7¾	2 10	3 0¼	3 2½	3 4⅞	3 9⅜	4 1⅞	4 6⅜	4 10⅞	5 3½

Table of the Diameter of Wheels at the Pitch Circle—Continued.

Number of Teeth.	Pitch of the Teeth.									
	inch. 1¾.	inch. 1⅞.	inches. 2.	inches. 2⅛.	inches. 2¼.	inches. 2½.	inches. 2¾.	inches. 3.	inches. 3¼.	inches. 3½.
58	2 8¼	2 10⅝	3 0⅞	3 3¼	3 5½	3 10⅛	4 2¾	4 7⅜	5 0	5 4⅝
59	2 8⅞	2 11¼	3 1½	3 4	3 6¼	3 11⅛	4 3⅝	4 8⅜	5 1	5 5¾
60	2 9⅜	2 11¾	3 2¼	3 4⅝	3 7	3 11¾	4 4½	4 9¼	5 2	5 6⅞
61	2 10	3 0⅜	3 2⅞	3 5¼	3 7¾	4 0½	4 5¼	4 10¼	5 3⅛	5 8
62	2 10½	3 1	3 3½	3 6	3 8½	4 1⅜	4 6¼	4 11¼	5 4⅛	5 9
63	2 11	3 1⅝	3 4⅛	3 6⅝	3 9⅛	4 2⅛	4 7⅛	5 0⅛	5 5⅛	5 10⅛
64	2 11⅝	3 2¼	3 4¾	3 7¼	3 9⅞	4 3	4 8	5 1⅛	5 6¼	5 11⅜
65	3 0¼	3 2⅞	3 5⅜	3 8	3 10½	4 3¾	4 8⅞	5 2	5 7¼	6 0⅜
66	3 0⅞	3 3⅜	3 6	3 8⅝	3 11¼	4 4½	4 9¾	5 3	5 8¼	6 1½
67	3 1⅜	3 4	3 6½	3 9⅜	4 0	4 5⅜	4 10⅝	5 4	5 9⅜	6 2⅝
68	3 1¾	3 4⅝	3 7¼	3 10	4 0¾	4 6⅛	4 11½	5 5	5 10⅜	6 3¾
69	3 2⅜	3 5⅛	3 7⅞	3 10⅝	4 1½	4 7	5 0⅜	5 6	5 11⅜	6 4⅞
70	3 3	3 5¾	3 8½	3 11⅜	4 2⅛	4 7¾	5 1¼	5 6⅞	6 0½	6 6
71	3 3½	3 6⅜	3 9¼	4 0	4 2⅞	4 8½	5 2⅛	5 7¾	6 1½	6 7
72	3 4⅛	3 6¾	3 9⅞	4 0¾	4 3½	4 9¼	5 3	5 8¾	6 2½	6 8¼
73	3 4⅝	3 7½	3 10½	4 1⅜	4 4¼	4 10	5 3⅞	5 9¾	6 3½	6 9⅜
74	3 5¼	3 7⅞	3 11⅛	4 2	4 5	4 10⅞	5 4¾	5 10⅝	6 4½	6 10⅝
75	3 5¾	3 8¾	3 11¾	4 2¾	4 5¾	4 11¾	5 5⅜	5 11⅝	6 5½	6 11½
76	3 6⅜	3 9⅜	4 0⅜	4 3½	4 6½	5 0½	5 6½	6 0½	6 6⅝	7 0⅝
77	3 6⅞	3 9⅞	4 1	4 4	4 7⅛	5 1¼	5 7⅜	6 1½	6 7⅝	7 1¼
78	3 7½	3 10½	4 1⅝	4 4¾	4 7⅞	5 2	5 8¼	6 2½	6 8⅝	7 2⅞
79	3 8	3 11⅛	4 2¼	4 5½	4 8½	5 2⅞	5 9⅛	6 3½	6 9¾	7 4
80	3 8½	3 11¾	4 3	4 6⅛	4 9¼	5 3⅝	5 10	6 4⅜	6 10¾	7 5⅛
81	3 9⅛	4 0⅜	4 3½	4 6⅞	4 10	5 4½	5 10⅞	6 5⅜	6 11⅞	7 6¼
82	3 9⅝	4 0⅞	4 4¼	4 7½	4 10¾	5 5¼	5 11¾	6 6⅜	7 0⅞	7 7⅜
83	3 10¼	4 1½	4 4⅞	4 8⅛	4 11½	5 6	6 0⅝	6 7¼	7 1⅞	7 8⅝
84	3 10¾	4 2⅛	4 5½	4 8⅞	5 0⅛	5 6⅞	6 1½	6 8⅛	7 2⅞	7 9½
85	3 11¼	4 2¾	4 6⅛	4 9½	5 0⅞	5 7⅝	6 2⅜	6 9⅛	7 3⅞	7 10⅝
86	3 11⅞	4 3¼	4 6¾	4 10⅛	5 1⅝	5 8½	6 3¼	6 10½	7 5	7 11¾
87	4 0½	4 3⅞	4 7⅜	4 10⅞	5 2¼	5 9¼	6 4⅛	6 11	7 6	8 0⅞
88	4 1	4 4½	4 8	4 11½	5 3	5 10	6 5	7 0	7 7	8 2
89	4 1½	4 5⅛	4 8⅝	5 0¼	5 3¼	5 10¾	6 5⅞	7 1	7 8	8 3⅛
90	4 2⅛	4 5¾	4 9¼	5 0⅞	5 4¼	5 11⅝	6 6¾	7 2	7 9⅛	8 4¼
91	4 2¾	4 6¼	4 9⅞	5 1½	5 5⅛	6 0⅜	6 7⅝	7 2⅞	7 10⅛	8 5⅜
92	4 3¼	4 7	4 10½	5 2⅛	5 5⅞	6 1	6 8½	7 3⅞	7 11⅛	8 6½
93	4 3⅞	4 7½	4 11¼	5 2⅞	5 6⅝	6 2	6 9⅜	7 4⅞	8 0¼	8 7⅝
94	4 4⅜	4 8⅛	4 11¾	5 3½	5 7⅜	6 2¾	6 10¼	7 5¾	8 1¼	8 8¾
95	4 4⅞	4 8¾	5 0½	5 4¼	5 8	6 3½	6 11⅛	7 6¾	8 2¼	8 9⅞
96	4 5½	4 9⅜	5 1⅛	5 5	5 8¾	6 4⅜	7 0	7 7⅝	8 3⅜	8 10⅞
97	4 6	4 10	5 1¾	5 5⅝	5 9½	6 5¼	7 0⅞	7 8⅝	8 4⅜	9 0
98	4 6½	4 10½	5 2⅜	5 6¼	5 10⅛	6 6	7 1¾	7 9½	8 5⅜	9 1⅛
99	4 7⅛	4 11	5 3	5 7	5 11	6 6¾	7 2⅝	7 10½	8 6½	9 2¼
100	4 7¾	4 11⅝	5 3⅝	5 7⅝	5 11⅝	6 7½	7 3⅝	7 11½	8 7½	9 3⅜
101	4 8¼	5 0¼	5 4¼	5 8¼	6 0¼	6 8⅜	7 4⅜	8 0½	8 8½	9 4½
102	4 8⅞	5 1	5 5	5 9	6 1	6 9⅛	7 5¼	8 1⅜	8 9½	9 5⅝
103	4 9⅜	5 1¼	5 5½	5 9⅝	6 1½	6 10	7 6⅛	8 2¼	8 10½	9 6¾
104	4 10	5 1¾	5 6¼	5 10¼	6 2¼	6 10¾	7 7	8 3¼	8 11½	9 7⅞

Table of the Diameter of Wheels at the Pitch Circle—Continued.

Number of Teeth.	Pitch of the Teeth.									
	inch. 1¾.	inch. 1⅞.	inches. 2.	inches. 2⅛.	inches. 2¼.	inches. 2½.	inches. 2¾.	inches. 3.	inches. 3¼.	inches. 3½.
105	4 10½	5 2⅜	5 6¾	5 11	6 3	6 11½	7 7⅞	8 4¼	9 0⅝	9 8⅞
106	4 11	5 2⅞	5 7½	5 11⅝	6 3⅝	7 0¼	7 8¾	8 5⅛	9 1⅝	9 10
107	4 11½	5 3½	5 8½	6 0⅜	6 4¼	7 1⅛	7 9⅝	8 6⅛	9 2⅝	9 11¼
108	5 0⅛	5 4⅛	5 8¾	6 1	6 5	7 2	7 10½	8 7⅛	9 3¾	10 0⅜
109	5 0¾	5 4¾	5 9⅜	6 1¾	6 5¾	7 2¾	7 11⅝	8 8	9 4¾	10 1⅜
110	5 1¼	5 5¼	5 10	6 2⅜	6 6½	7 3½	8 0¼	8 9	9 5¾	10 2⅝
111	5 1⅞	5 5⅞	5 10⅝	6 3	6 7⅛	7 4¼	8 1⅛	8 10	9 6⅞	10 3⅝
112	5 2⅜	5 6½	5 11¼	6 3¾	6 8	7 5⅛	8 2	8 10⅞	9 7⅞	10 4¾
113	5 3	5 7	6 0	6 4⅝	6 8⅝	7 6	8 3	8 11⅞	9 8⅞	10 6
114	5 3½	5 7⅝	6 0½	6 5⅛	6 9¼	7 6¾	8 3¾	9 0¾	9 9⅞	10 7
115	5 4	5 8¼	6 1¼	6 5¾	6 10	7 7½	8 4⅝	9 1¾	9 10⅞	10 8⅛
116	5 4⅝	5 8⅞	6 1¾	6 6½	6 10¾	7 8¼	8 5½	9 2¾	10 0	10 9¼
117	5 5⅛	5 9½	6 2½	6 7⅛	6 11½	7 9⅛	8 6⅜	9 3⅝	10 1	10 10⅜
118	5 5¾	5 10	6 3⅛	6 7¾	7 0⅛	7 10	8 7¼	9 4⅝	10 2	10 11½
119	5 6¼	5 10⅝	6 3¾	6 8½	7 1	7 10¾	8 8⅛	9 5½	10 3⅛	11 0½
120	5 6¾	5 11¼	6 4⅜	6 9⅛	7 1⅝	7 11½	8 9	9 6½	10 4⅛	11 1⅝
121	5 7⅜	5 11⅞	6 5	6 9¾	7 2¼	8 0¼	8 9⅞	9 7½	10 5⅛	11 2¾
122	5 8	6 0½	6 5⅝	6 10½	7 3	8 1	8 10¾	9 8½	10 6¼	11 3⅞
123	5 8½	6 1	6 6¼	6 11⅛	7 3¾	8 1⅞	8 11⅝	9 9⅜	10 7¼	11 5
124	5 9	6 1⅝	6 7	6 11⅞	7 4½	8 2⅝	9 0½	9 10⅜	10 8¼	11 6⅛
125	5 9⅝	6 2¼	6 7½	7 0½	7 5⅛	8 3½	9 1⅜	9 11¼	10 9⅜	11 7¼
126	5 10⅛	6 2¾	6 8¼	7 1¼	7 6	8 4¼	9 2¼	10 0¼	10 10⅜	11 8⅜
127	5 10¾	6 3⅜	6 8¾	7 2	7 6⅝	8 5	9 3⅛	10 1¼	10 11⅜	11 9½
128	5 11¼	6 4	6 9½	7 2½	7 7¼	8 5⅞	9 4	10 2¼	11 0½	11 10½
129	5 11⅞	6 4½	6 10⅛	7 3¼	7 8	8 6⅝	9 4⅞	10 3⅛	11 1½	11 11¾
130	6 0⅜	6 5⅛	6 10¾	7 4	7 8¾	8 7½	9 5¾	10 4⅛	11 2½	12 0⅞
131	6 1	6 5¾	6 11⅜	7 4½	7 9½	8 8¼	9 6⅝	10 5	11 3½	12 2
132	6 1½	6 6⅜	7 0	7 5⅛	7 10⅛	8 9	9 7½	10 6	11 4⅝	12 3
133	6 2	6 7	7 0⅝	7 6	7 10⅞	8 9¾	9 8⅜	10 7	11 5⅝	12 4⅛
134	6 2⅝	6 7½	7 1¼	7 6⅝	7 11½	8 10⅝	9 9¼	10 8	11 6⅝	12 5¼
135	6 3⅛	6 8½	7 2	7 7¼	8 0¼	8 11⅜	9 10⅛	10 8⅞	11 7⅝	12 6⅜
136	6 3¾	6 8¾	7 2½	7 8	8 1	9 0¼	9 11	10 9¾	11 8⅝	12 7½
137	6 4¼	6 9¼	7 3¼	7 8⅝	8 1¾	9 1	10 0	10 10¾	11 9¾	12 8⅝
138	6 4¾	6 10	7 3¾	7 9¼	8 2½	9 1¾	10 0⅞	10 11¾	11 10¾	12 9¾
139	6 5½	6 10½	7 4½	7 10	8 3¼	9 2⅝	10 1⅝	11 0⅝	11 11¾	12 10⅞
140	6 6	6 11⅛	7 5⅛	7 10⅝	8 3⅞	9 3⅜	10 2½	11 1⅝	12 0⅞	13 0
141	6 6½	6 11¾	7 5¾	7 11⅜	8 4½	9 4¼	10 3⅜	11 2½	12 1⅞	13 1
142	6 7	7 0¼	7 6¾	8 0	8 5¼	9 5	10 4¼	11 3½	12 2⅞	13 2⅛
143	6 7⅝	7 0⅞	7 7	8 0¾	8 6	9 5¾	10 5⅛	11 4½	12 3⅞	13 3¼
144	6 8¼	7 1½	7 7⅝	8 1⅜	8 6¾	9 6	10 6	11 5½	12 4⅞	13 4⅜
145	6 8¾	7 2	7 8¼	8 2	8 7½	9 7⅜	10 6⅞	11 6½	12 6	13 5½
146	6 9¼	7 2⅝	7 9	8 2¾	8 8½	9 8⅛	10 7¾	11 7⅜	12 7	13 6⅝
147	6 9⅞	7 3¼	7 9½	8 3⅜	8 8⅞	9 9	10 8⅝	11 8¼	12 8	13 7¾
148	6 10½	7 3⅞	7 10¼	8 4⅛	8 9½	9 9¾	10 9½	11 9¼	12 9⅛	13 8⅞
149	6 11	7 4½	7 10¾	8 4⅞	8 10¼	9 10½	10 10⅜	11 10¼	12 10⅛	13 10
150	6 11½	7 5	7 11½	8 5¼	8 11	9 11⅜	10 11¼	11 11¼	12 11⅛	13 11⅛

Table of the Diameter of Wheels at the Pitch Circle—Continued.

Number of Teeth.	Pitch of the Teeth.					
	inches. 2¼.	inches. 2½.	inches. 2¾.	inches. 3.	inches. 3¼.	inches. 3½.
151	9 0⅛	10 0⅛	11 0⅛	12 0⅛	13 0⅛	14 0¼
152	9 0⅞	10 0⅞	11 1	12 1⅛	13 1¼	14 1⅜
153	9 1⅝	10 1¾	11 1⅞	12 2	13 2¼	14 2⅜
154	9 2¼	10 2½	11 2⅞	12 3	13 3⅜	14 3½
155	9 3	10 3⅜	11 3⅝	12 4	13 4⅜	14 4⅝
156	9 3⅝	10 4⅛	11 4½	12 4⅞	13 5⅜	14 5¾
157	9 4⅜	10 4⅞	11 5⅝	12 5⅞	13 6⅜	14 6⅞
158	9 5⅛	10 5⅝	11 6⅜	12 6⅞	13 7⅜	14 8
159	9 5⅞	10 6½	11 7⅛	12 7⅞	13 8⅜	14 9⅛
160	9 6½	10 7⅜	11 8	12 8¾	13 9½	14 10¼
161	9 7⅜	10 8⅛	11 8⅞	12 9⅝	13 10½	14 11⅜
162	9 8	10 8⅞	11 9¾	12 10⅝	13 11½	15 0⅜
163	9 8¾	10 9⅝	11 10⅝	12 11⅝	14 0¾	15 1½
164	9 9⅝	10 10½	11 11½	13 0⅝	14 1⅝	15 2⅝
165	9 10⅛	10 11⅜	12 0¾	13 1½	14 2⅜	15 3¾
166	9 10⅞	11 0	12 1¼	13 2½	14 3¾	15 4⅞
167	9 11⅝	11 0⅞	12 2⅛	13 3⅜	14 4⅝	15 6
168	10 0⅜	11 1⅝	12 3	13 4⅜	14 5¾	15 7⅛
169	10 1	11 2⅜	12 3⅞	13 5⅜	14 6⅝	15 8¼
170	10 1¾	11 3¼	12 4¾	13 6⅜	14 7⅝	15 9⅝
171	10 2⅜	11 4	12 5⅝	13 7¼	14 8⅞	15 10½
172	10 3⅛	11 4⅞	12 6½	13 8⅛	14 9⅞	15 11⅝
173	10 3⅞	11 5¾	12 7⅜	13 9⅛	14 10⅞	16 0⅝
174	10 4½	11 6⅝	12 8¼	13 10⅛	15 0	16 1¾
175	10 5⅜	11 7¼	12 9⅛	13 11⅛	15 1⅜	16 2⅞
176	10 6	11 8	12 10	14 0	15 2⅝	16 4
177	10 6¾	11 8⅞	12 10⅞	14 1	15 3⅛	16 5⅛
178	10 7⅝	11 9⅝	12 11⅞	14 1⅞	15 4¼	16 6¾
179	10 8¼	11 10⅜	13 0⅝	14 2⅞	15 5⅛	16 7⅜
180	10 8⅞	11 11¼	13 1½	14 3⅞	15 6¼	16 8½
181	10 9⅝	12 0	13 2⅝	14 4⅞	15 7¼	16 9⅝
182	10 10⅜	12 0⅞	13 3⅜	14 5¾	15 8¼	16 10¾
183	10 11	12 1⅝	13 4⅛	14 6¾	15 9⅜	16 11⅞
184	10 11¾	12 2⅜	13 5	14 7⅝	15 10⅜	17 0⅞
185	11 0⅜	12 3⅛	13 5⅞	14 8⅝	15 11⅜	17 2⅛
186	11 1⅛	12 4	13 6⅞	14 9⅝	16 0⅜	17 3¼
187	11 1⅞	12 4⅞	13 7⅝	14 10½	16 1⅜	17 4⅜
188	11 2⅝	12 5½	13 8½	14 11½	16 2⅜	17 5⅜
189	11 3⅜	12 6⅜	13 9⅜	15 0⅜	16 3½	17 6½
190	11 4	12 7⅛	13 10⅜	15 1⅜	16 4½	17 7⅝
191	11 4¾	12 7⅞	13 11⅛	15 2⅜	16 5½	17 8¾
192	11 5½	12 8⅝	14 0	15 3⅜	16 6⅝	17 9⅞
193	11 6⅛	12 9½	14 0⅞	15 4¼	16 7⅝	17 11
194	11 6⅞	12 10⅜	14 1⅞	15 5¼	16 8⅝	18 0⅛
195	11 7⅝	12 11⅛	14 2⅝	15 6⅛	16 9¾	18 1¼

Table of the Diameter of Wheels at the Pitch Circle—Continued.

Number of Teeth.	Pitch of the Teeth.					
	inches. 2¼.	inches. 2½.	inches. 2¾.	inches. 3.	inches. 3¼.	inches. 3½.
196	11 8⅜	12 11⅞	14 3½	15 7⅛	16 10¾	18 2⅜
197	11 9	13 0¾	14 4⅜	15 8⅛	16 11¾	18 3½
198	11 9⅞	13 1½	14 5⅜	15 9	17 0⅞	18 4½
199	11 10½	13 2⅜	14 6⅛	15 10	17 1⅞	18 5⅝
200	11 11⅛	13 3⅛	14 7	15 10⅞	17 2⅞	18 6¾
201	11 11⅞	13 3⅞	14 7⅞	15 11⅞	17 3⅞	18 7⅞
202	12 0⅝	13 4¾	14 8⅞	16 0⅞	17 4⅞	18 9
203	12 1⅜	13 5⅝	14 9⅝	16 1⅞	17 6	18 10
204	12 2	13 6¼	14 10½	16 2⅞	17 7	18 11⅛
205	12 2⅞	13 7⅛	14 11⅜	16 3¾	17 8	19 0¼
206	12 3½	13 7⅞	15 0⅜	16 4⅝	17 9⅛	19 1⅝
207	12 4¼	13 8⅝	15 1⅛	16 5⅝	17 10⅛	19 2⅝
208	12 4⅞	13 9⅜	15 2	16 6⅝	17 11⅛	19 3⅝
209	12 5⅝	13 10¼	15 2⅞	16 7½	18 0¼	19 4¾
210	12 6⅜	13 11⅛	15 3⅞	16 8½	18 1¼	19 5⅞
211	12 7⅛	13 11⅞	15 4⅝	16 9⅜	18 2¼	19 7
212	12 7⅞	14 0⅝	15 5½	16 10⅜	18 3⅜	19 8⅛
213	12 8½	14 1⅝	15 6⅜	16 11⅜	18 4⅜	19 9¼
214	12 9¼	14 2⅛	15 7⅜	17 0⅜	18 5⅜	19 10⅜
215	12 9⅞	14 3	15 8⅛	17 1⅜	18 6⅜	19 11½
216	12 10⅝	14 3⅞	15 9	17 2¼	18 7⅜	20 0⅝
217	12 11⅜	14 4⅝	15 9⅞	17 3⅛	18 8⅜	20 1¾
218	13 0	14 5⅜	15 10⅞	17 4⅛	18 9½	20 2⅞
219	13 0⅞	14 6¼	15 11⅝	17 5⅛	18 10½	20 4
220	13 1½	14 7	16 0½	17 6	18 11½	20 5⅛
221	13 2¼	14 7⅞	16 1⅜	17 7	19 0⅝	20 6¼
222	13 2⅞	14 8⅝	16 2⅜	17 7⅞	19 1⅝	20 7⅜
223	13 3⅝	14 9⅜	16 3⅛	17 8⅞	19 2⅞	20 8½
224	13 4⅜	14 10¼	16 4	17 9⅞	19 3¾	20 9⅝
225	13 5⅛	14 11	16 4⅞	17 10⅞	19 4¾	20 10¾
226	13 5⅞	14 11⅞	16 5⅞	17 11⅞	19 5¾	20 11⅞
227	13 6½	15 0⅝	16 6⅝	18 0¾	19 6⅞	21 0⅞
228	13 7¼	15 1⅜	16 7½	18 1⅝	19 7⅞	21 2
229	13 8	15 2⅛	16 8⅜	18 2⅝	19 8⅞	21 3⅛
230	13 8⅝	15 3	16 9⅜	18 3⅝	19 9⅞	21 4¼
231	13 9⅜	15 3⅞	16 10⅛	18 4½	19 10⅞	21 5⅜
232	13 10⅛	15 4⅝	16 11	18 5½	20 0	21 6⅜
233	13 10⅞	15 5⅜	16 11⅞	18 6½	20 1	21 7½
234	13 11½	15 6⅛	17 0⅞	18 7⅜	20 2	21 8⅝
235	14 0¼	15 7	17 1⅝	18 8⅜	20 3	21 9⅞
236	14 1	15 7⅞	17 2½	18 9⅜	20 4⅛	21 10⅞
237	14 1⅝	15 8½	17 3⅜	18 10⅜	20 5⅛	22 0
238	14 2⅜	15 9⅜	17 4⅜	18 11¼	20 6¼	22 1⅛
239	14 3⅛	15 10⅛	17 5⅛	19 0⅛	20 7¼	22 2¼
240	14 3⅞	15 10⅞	17 6	19 1⅛	20 8¼	22 3⅜

Table of the Diameter of Wheels at the Pitch Circle—Continued.

Number of Teeth.	Pitch of the Teeth. inches. 2¼	inches. 2½.	inches. 2¾.	inches. 3.	inches. 3¼.	inches. 3½.
241	14 4½	15 11⅞	17 6⅞	19 2⅞	20 9⅜	22 4⅜
242	14 5¼	16 0½	17 7⅝	19 3⅞	20 10⅜	22 5⅝
243	14 6	16 1⅜	17 8⅝	19 4⅝	20 11⅜	22 6⅝
244	14 6¾	16 2⅛	17 9½	19 5½	21 0⅜	22 7⅞
245	14 7⅜	16 2⅞	17 10¾	19 6⅜	21 1⅜	22 8⅞
246	14 8⅛	16 3¾	17 11⅜	19 7⅜	21 2⅜	22 10
247	14 8⅞	16 4½	18 0⅛	19 8⅛	21 3½	22 11⅛
248	14 9⅝	16 5⅛	18 1	19 9	21 4½	23 0¼
249	14 10⅜	16 6⅛	18 1⅞	19 9⅞	21 5½	23 1⅛
250	14 11	16 6⅞	18 2⅞	19 10⅞	21 6⅝	23 2⅛
251	14 11¾	16 7⅝	18 3⅝	19 11⅝	21 7⅝	23 3⅜
252	15 0⅜	16 8½	18 4⅝	20 0⅝	21 8⅝	23 4¾
253	15 1⅛	16 9¼	18 5⅝	20 1½	21 9⅝	23 5⅞
254	15 1⅞	16 10⅛	18 6⅜	20 2½	21 10¾	23 6⅞
255	15 2⅝	16 10⅞	18 7⅜	20 3½	21 11¾	23 8
256	15 3⅜	16 11⅞	18 8	20 4⅜	22 0⅞	23 9⅛
257	15 4	17 0½	18 8⅞	20 5⅜	22 1⅞	23 10⅜
258	15 4¾	17 1⅜	18 9⅞	20 6⅜	22 2⅞	23 11⅜
259	15 5⅜	17 2⅛	18 10¼	20 7⅜	22 3⅞	24 0½
260	15 6¼	17 2⅞	18 11½	20 8¼	22 4⅞	24 1⅝
261	15 6⅞	17 3⅝	19 0⅜	20 9⅛	22 6	24 2¾
262	15 7⅝	17 4⅜	19 1⅜	20 10⅛	22 7	24 3⅞
263	15 8⅜	17 5¼	19 2⅛	20 11⅛	22 8	24 5
264	15 9	17 6	19 3	21 0⅛	22 9⅛	24 6⅛
265	15 9¾	17 6⅞	19 3⅞	21 1	22 10⅛	24 7¼
266	15 10⅝	17 7⅝	19 4⅞	21 2	22 11⅛	24 8⅜
267	15 11⅛	17 8⅜	19 5⅝	21 2⅞	23 0¼	24 9⅜
268	15 11⅞	17 9¼	19 6½	21 3⅞	23 1¼	24 10½
269	16 0⅝	17 10	19 7⅜	21 4⅞	23 2¼	24 11⅝
270	16 1⅜	17 10⅞	19 8⅜	21 5⅞	23 3⅜	25 0⅞
271	16 2	17 11⅝	19 9⅜	21 6¾	23 4⅜	25 1⅞
272	16 2¾	18 0⅜	19 10	21 7⅝	23 5⅜	25 3
273	16 3½	18 1⅛	19 10⅞	21 8⅝	23 6⅜	25 4⅛
274	16 4⅛	18 2	19 11⅞	21 9⅝	23 7⅜	25 5¼
275	16 4⅞	18 2⅞	20 0⅝	21 10⅝	23 8½	25 6⅜
276	16 5⅝	18 3⅝	20 1½	21 11½	23 9½	25 7⅜
277	16 6⅜	18 4⅜	20 2⅜	22 0½	23 10½	25 8⅝
278	16 7	18 5¼	20 3⅜	22 1⅜	25 11½	25 9⅝
279	16 7⅞	18 6	20 4⅛	22 2⅜	24 0⅝	25 10⅞
280	16 8½	18 6⅞	20 5	22 3⅜	24 1⅝	25 11⅞
281	16 9¼	18 7⅝	20 5⅞	22 4⅜	24 2⅝	26 1
282	16 9⅞	18 8⅜	20 6⅞	22 5¼	24 3¾	26 2⅛
283	16 10⅝	18 9⅛	20 7⅝	22 6⅛	24 4¾	26 3¼
284	16 11⅜	18 9⅞	20 8½	22 7⅛	24 5¾	26 4⅜
285	17 0⅛	18 10¾	20 9⅜	22 8⅛	24 6⅞	26 5½

Table of the Diameter of Wheels at the Pitch Circle—Continued.

Number of Teeth.	Pitch of the Teeth.					
	inches. 2¼.	inches. 2½.	inches. 2¾.	inches. 3.	inches. 3¼.	inches. 3½.
286	17 0⅞	18 11½	20 10⅜	22 9	24 7⅞	26 6⅝
287	17 1½	19 0⅜	20 11⅝	22 10	24 8⅞	26 7⅝
288	17 2¼	19 1⅛	21 0	22 11	24 9⅞	26 8⅞
289	17 2⅞	19 1⅞	21 0⅞	22 11⅞	24 10⅞	26 9⅞
290	17 3⅝	19 2⅝	21 1⅞	23 0⅞	25 0	26 11
291	17 4⅜	19 3½	21 2¾	23 1⅞	25 1	27 0⅛
292	17 5⅛	19 4½	21 3⅝	23 2⅞	25 2	27 1⅜
293	17 5⅞	19 5⅛	21 4⅜	23 3¾	25 3⅛	27 2⅜
294	17 6½	19 5⅞	21 5¼	23 4⅝	25 4⅛	27 3½
295	17 7¼	19 6⅜	21 6⅛	23 5⅝	25 5⅛	27 4⅝
296	17 7⅞	19 7½	21 7	23 6⅝	25 6¼	27 5¾
297	17 8⅝	19 8⅜	21 7⅞	23 7⅝	25 7¼	27 6⅞
298	17 9⅜	19 9⅛	21 8⅞	23 8½	25 8¼	27 7⅞
299	17 10⅛	19 9⅞	21 9⅝	23 9½	25 9⅜	27 9⅛
300	17 10⅞	19 10⅝	21 10⅝	23 10⅝	25 10⅜	27 10¼

Weight of Cast Iron Balls from 1 to 12 Inches Diameter.

Size.	Wt.	Size.	Wt.	Size.	Wt.	Size.	Wt.	Size.	Wt.
Inch.	lbs.	Inch.	lbs.	Inch.	lbs.	Inch.	lbs.	Inch.	lbs.
1	.136	3½	5.84	6	29.45	8½	83.73	11	181.48
1½	.460	4	8.72	6½	37.44	9	99.4	11½	207.37
2	1.09	4½	12.42	7	46.76	9½	116.9	12	235.62
2½	2.13	5	17.04	7½	57.52	10	136.35		
3	3.68	5½	22.68	8	69.81	10½	157.84		

Weight of Cast Iron Pipes 12 Inches Long, from ¼ to 1¼ Inch Thick.

Diam. of Bore.	Inch. ¼	Inch ⅜	Inch. ½	Inch. ⅝	Inch. ¾	Inch. ⅞	Inch. 1	Inch. 1⅛	Inch. 1¼
Inch.	lbs.	lbs.	lbs.	lbs.	lbs.	lbs.	lbs.	lbs.	lbs.
1	3.06	5.06	7.36	9.97	12.89	16.11	19.63		
1¼	3.68	5.98	8.59	11.51	14.73	18.25	22.09		
1½	4.29	6.9	9.82	13.04	16.56	20.4	24.54	28.99	33.74
1¾	4.91	7.83	11.05	14.57	18.41	22.55	27.	31.75	36.76
2	5.53	8.75	12.27	16.11	20.25	24.7	29.45	34.46	39.89
2¼	6.14	9.66	13.5	17.64	22.09	26.84	31.85	37.28	42.95
2½	6.74	10.58	14.72	19.17	23.92	28.93	34.36	40.03	46.02
2¾	7.36	11.5	15.95	20.7	25.71	31.14	36.81	42.8	49.08
3	7.98	12.43	17.18	22.19	27.62	33.29	39.28	45.56	52.16
3¼	8.59	13.34	18.35	23.78	29.45	35.44	41.72	48.32	55.22
3½	9.2	14.21	19.64	25.31	31.3	37.58	44.18	51.08	58.29
3¾	9.76	15.19	20.86	26.85	33.13	39.73	46.63	53.84	61.36
4	10.44	16.11	22.1	28.38	34.98	41.88	49.09	56.61	64.43
4¼	11.1	17.08	23.37	29.97	36.87	44.08	51.6	59.42	67.55
4½	11.66	17.94	24.54	31.44	38.65	46.17	53.99	62.12	70.56
4¾	12.27	18.87	25.77	32.98	40.5	48.32	56.45	64.89	73.63
5	12.88	19.78	26.99	34.51	42.33	50.46	58.9	67.64	76.69
5¼	13.5	20.71	28.23	36.05	44.18	52.62	61.36	70.41	79.77
5½	14.11	21.63	29.45	37.58	46.02	54.76	63.81	73.17	82.84
5¾	14.73	22.55	30.68	39.12	47.86	56.91	66.27	75.94	85.91

Weight of Cast Iron Pipes 12 Inches Long, from ¼ to 1¼ Inch Thick—Cont.

Diam. of Bore.	Inch, ¼	Inch, ⅜	Inch, ½	Inch, ⅝	Inch, ¾	Inch, ⅞	Inch, 1	Inch, 1⅛	Inch, 1¼
Inch.	lbs.	lbs.	lbs.	lbs.	lbs.	lbs.	lbs.	lbs.	lbs.
6	15.34	23.47	31.91	40.65	49.7	59.06	68.73	78.7	88.75
6¼	15.95	24.39	33.13	42.18	51.54	61.21	71.18	81.23	92.04
6½	16.57	25.31	34.36	43.72	53.39	63.36	73.41	84.22	95.1
6¾	17.18	26.23	35.59	45.26	55.23	65.28	76.09	86.97	98.18
7	17.79	27.15	36.82	46.79	56.84	67.65	78.53	89.74	101.24
7¼	18.41	28.08	38.05	48.1	58.91	69.79	81.	92.5	104.31
7½	19.03	29.	39.05	49.86	60.74	71.95	83.45	95.26	107.38
7¾	19.64	29.69	40.5	51.38	62.59	74.09	85.9	98.02	110.45
8	20.02	30.83	41.71	52.92	64.42	76.23	88.35	100.78	113.51
8¼	20.86	31.74	42.95	54.45	66.26	78.38	90.81	103.54	116.58
8½	21.69	32.9	44.4	56.21	68.33	80.76	93.49	106.53	119.87
8¾	22.09	33.59	45.4	57.52	69.95	82.68	95.72	109.06	122.72
9	22.71	34.52	46.64	59.07	71.8	84.84	98.18	111.84	125.8
9¼	23.31	35.43	47.86	60.59	73.63	86.97	100.63	114.59	128.85
9½	23.93	36.36	49.09	62.13	75.47	89.13	103.09	117.35	131.93
9¾	24.55	37.28	50.32	63.66	77.32	91.28	105.54	120.12	134.99
10	25.16	38.2	51.54	65.2	79.16	93.42	108.	122.87	138.06
10¼	25.77	39.11	52.77	66.73	80.99	95.57	110.44	125.63	141.12
10½	26.38	40.04	54.	68.26	82.84	97.71	112.9	128.39	144.19
10¾	27.	40.96	55.22	69.8	84.67	99.86	115.35	131.15	147.26
11	27.62	41.88	56.46	71.33	86.52	102.01	117.81	133.92	150.33
11¼	28.22	42.8	57.67	72.86	88.35	104.15	120.26	136.67	153.4
11½	28.84	43.71	58.9	74.39	90.19	106.3	122.71	139.44	156.44
11¾	29.45	44.64	60.13	75.93	92.04	108.45	125.18	142.18	159.54
12	30.06	45.55	61.35	77.46	93.6	110.6	127.6	144.96	162.6

Weight of Cast Iron Pipes 12 Inches Long, from 1⅜ to 1½ Inch Thick.

D. of B.	1⅜ Inch.	1½ Inch.	D. of B.	1⅜ Inch.	1½ Inch.	D. of B.	1⅜ Inch.	1½ Inch.
Inch.	lbs.	lbs.	Inch.	lbs.	lbs.	Inch.	lbs.	lbs.
2¼	48.94	55.22	5¾	95.96	106.77	9	140.06	154.64
2½	52.30	58.9	6	99.56	110.44	9¼	143.43	158.3
2¾	55.68	62.58	6¼	102.92	114.13	9½	146.8	161.99
3	59.06	66.27	6½	106.31	117.81	9¾	150.18	165.67
3¼	62.43	69.95	6¾	109.68	121.49	10	153.55	169.35
3½	65.81	73.63	7	113.05	125.17	10¼	156.92	173.03
3¾	69.18	77.31	7¼	116.43	128.86	10½	160.3	176.71
4	72.56	81.	7½	119.81	132.54	10¾	163.67	180.4
4¼	75.99	84.73	7¾	123.18	136.22	11	167.06	184.06
4½	79.3	88.35	8	126.55	139.89	11¼	170.4	187.76
4¾	82.68	92.04	8¼	129.92	143.58	11½	173.8	191.44
5	86.05	95.72	8½	133.53	147.49	11¾	177.18	195.12
5¼	89.44	99.41	8¾	136.68	150.94	12	180.54	198.8
5½	92.81	102.86						

Round Cast Iron Twelve Inches Long.

Size.	Weight.	Size.	Weight.	Size.	Weight.	Size.	Weight.	Size.	Weight.
Inch.	lbs.	Inch.	lbs.	Inch.	lbs.	Inch.	lbs.	Inch.	lbs.
$\frac{1}{4}$	.61	$2\frac{1}{4}$	12.42	4	39.27	$5\frac{3}{4}$	81.14	9	198.79
$\frac{1}{2}$	.95	$2\frac{3}{8}$	13.84	$4\frac{1}{8}$	41.76	$5\frac{7}{8}$	84.71	$9\frac{1}{8}$	210.
$\frac{3}{4}$	1.38	$2\frac{1}{2}$	15.33	$4\frac{1}{4}$	44.27	6	88.35	$9\frac{1}{2}$	221.5
$\frac{7}{8}$	1.87	$2\frac{5}{8}$	16.91	$4\frac{3}{8}$	46.97	$6\frac{1}{8}$	95.87	$9\frac{3}{4}$	233.34
1	2.45	$2\frac{3}{4}$	18.56	$4\frac{1}{2}$	49.7	$6\frac{1}{2}$	103.69	10	245.43
$1\frac{1}{8}$	3.1	$2\frac{7}{8}$	20.28	$4\frac{5}{8}$	52.5	$6\frac{3}{4}$	111.82	$10\frac{1}{4}$	257.86
$1\frac{1}{4}$	3.83	3	22.08	$4\frac{3}{4}$	55.37	7	120.26	$10\frac{1}{2}$	270.59
$1\frac{3}{8}$	4.64	$3\frac{1}{8}$	23.96	$4\frac{7}{8}$	58.32	$7\frac{1}{4}$	129.	$10\frac{3}{4}$	283.63
$1\frac{1}{2}$	5.52	$3\frac{1}{4}$	25.92	5	61.35	$7\frac{1}{2}$	138.05	11	296.97
$1\frac{5}{8}$	6.48	$3\frac{3}{8}$	27.95	$5\frac{1}{8}$	64.46	$7\frac{3}{4}$	147.41	$11\frac{1}{4}$	310.63
$1\frac{3}{4}$	7.51	$3\frac{1}{2}$	30.06	$5\frac{1}{4}$	67.64	8	157.08	$11\frac{1}{2}$	324.59
$1\frac{7}{8}$	8.62	$3\frac{5}{8}$	32.25	$5\frac{3}{8}$	70.09	$8\frac{1}{4}$	167.05	$11\frac{3}{4}$	338.85
2	9.81	$3\frac{3}{4}$	34.51	$5\frac{1}{2}$	74.24	$8\frac{1}{2}$	177.1	12	353.43
$2\frac{1}{8}$	11.08	$3\frac{7}{8}$	36.85	$5\frac{5}{8}$	77.65	$8\frac{3}{4}$	187.91		

Square Cast Iron Twelve Inches Long.

Size.	Weight.	Size.	Weight.	Size.	Weight.	Size.	Weight.	Size.	Weight.
Inch.	lbs.	Inch.	lbs.	Inch.	lbs.	Inch.	lbs.	Inch.	lbs.
$\frac{1}{4}$	.78	$2\frac{1}{4}$	15.81	4	50.	$5\frac{3}{4}$	103.32	9	253.12
$\frac{1}{2}$	1.22	$2\frac{3}{8}$	17.62	$4\frac{1}{8}$	53.14	$5\frac{7}{8}$	107.86	$9\frac{1}{4}$	267.38
$\frac{3}{4}$	1.75	$2\frac{1}{2}$	19.53	$4\frac{1}{4}$	56.44	6	112.5	$9\frac{1}{2}$	282.
$\frac{7}{8}$	2.39	$2\frac{5}{8}$	21.53	$4\frac{3}{8}$	59.81	$6\frac{1}{8}$	122.08	$9\frac{3}{4}$	297.07
1	3.12	$2\frac{3}{4}$	23.63	$4\frac{1}{2}$	63.28	$6\frac{1}{2}$	132.03	10	312.5
$1\frac{1}{8}$	3.95	$2\frac{7}{8}$	25.83	$4\frac{5}{8}$	66.84	$6\frac{3}{4}$	142.38	$10\frac{1}{4}$	328.32
$1\frac{1}{4}$	4.88	3	28.12	$4\frac{3}{4}$	70.5	7	153.12	$10\frac{1}{2}$	344.53
$1\frac{3}{8}$	5.9	$3\frac{1}{8}$	30.51	$4\frac{7}{8}$	74.26	$7\frac{1}{4}$	164.25	$10\frac{3}{4}$	361.13
$1\frac{1}{2}$	7.03	$3\frac{1}{4}$	33.	5	78.12	$7\frac{1}{2}$	175.78	11	378.12
$1\frac{5}{8}$	8.25	$3\frac{3}{8}$	35.59	$5\frac{1}{8}$	82.08	$7\frac{3}{4}$	187.68	$11\frac{1}{4}$	395.5
$1\frac{3}{4}$	9.57	$3\frac{1}{2}$	38.28	$5\frac{1}{4}$	86.13	8	200.	$11\frac{1}{2}$	413.28
$1\frac{7}{8}$	10.98	$3\frac{5}{8}$	41.06	$5\frac{3}{8}$	90.28	$8\frac{1}{4}$	212.56	$11\frac{3}{4}$	431.44
2	12.5	$3\frac{3}{4}$	43.94	$5\frac{1}{2}$	94.53	$8\frac{1}{2}$	225.78	12	450.
$2\frac{1}{8}$	14.11	$3\frac{7}{8}$	46.92	$5\frac{5}{8}$	98.87	$8\frac{3}{4}$	239.25		

Flat Cast Iron Twelve Inches Long, ¼ to 1 Inch Thick.

Width of Iron.	Inch. ¼	Inch. ⅜	Inch. ½	Inch. ⅝	Inch. ¾	Inch. ⅞	Inch. 1
Inch.	lbs.	lbs.	lbs.	lbs.	lbs.	lbs.	lbs.
2	1.56	2.34	3.12	3.9	4.68	5.46	6.25
2¼	1.75	2.63	3.51	4.39	5.27	6.15	7.03
2½	1.95	2.92	3.9	4.88	5.85	6.83	7.81
2¾	2.14	3.22	4.29	5.37	6.44	7.51	8.59
3	2.34	3.51	4.68	5.85	7.03	8.2	9.37
3¼	2.53	3.8	5.07	6.34	7.61	8.88	10.15
3½	2.73	4.1	5.46	6.83	8.2	9.57	10.93
3¾	2.93	4.39	5.85	7.32	8.78	10.25	11.71
4	3.12	4.68	6.25	7.81	9.37	10.93	12.5
4¼	3.32	4.97	6.64	8.3	9.96	11.62	13.28
4½	3.51	5.27	7.03	8.78	10.54	12.3	14.06
4¾	3.71	5.56	7.42	9.27	11.13	12.98	14.84
5	3.9	5.86	7.81	9.76	11.71	13.67	15.62
5¼	4.1	6.15	8.2	10.25	12.3	14.35	16.4
5½	4.29	6.44	8.59	10.74	12.89	15.03	17.18
5¾	4.49	6.73	8.98	11.23	13.46	15.72	17.96
6	4.68	7.03	9.37	11.71	14.06	16.4	18.75

Weight of a Superficial Foot of Cast Iron from ¼ to 2 Inches Thick.

Thickness.	¼	⅜	½	⅝	¾	⅞	1	1⅛
Wt.	lbs. 9.37	lbs. 14.06	lbs. 18.75	lbs. 23.43	lbs. 28.12	lbs. 32.81	lbs. 37.5	lbs. 42.18

Thickness.	1¼	1⅜	1½	1⅝	1¾	1⅞	2	
Wt.	lbs. 46.87	lbs. 51.56	lbs. 56.25	lbs. 60.93	lbs. 65.62	lbs. 70.31	lbs. 75.	

Weight of Square Lead Twelve Inches Long, from 1 to 3 Inches Square.

Size.	1 in.	1⅛	1¼	1⅜	1½	1⅝	1¾	1⅞	2
Wt.	lbs, 4.93	lbs. 6.25	lbs. 7.71	lbs. 9.33	lbs. 11.11	lbs. 13.04	lbs. 15.12	lbs. 17.36	lbs. 19.75

Size.	2⅛	2¼	2⅜	2½	2⅝	2¾	2⅞	3	
Wt.	lbs. 22.29	lbs. 25.	lbs. 27.8	lbs. 30.86	lbs. 34.02	lbs. 37.34	lbs. 40.81	lbs. 44.44	

Weight of Round Lead Twelve Inches Long, from 1 to 3 Inches Diameter.

Size.	1 in.	1⅛	1¼	1⅜	1½	1⅝	1¾	1⅞	2
Wt.	lbs. 3.87	lbs. 4.9	lbs. 6.06	lbs. 7.33	lbs. 8.72	lbs. 10.24	lbs. 11.87	lbs. 13.63	lbs. 15.51

Size.	2⅛	2¼	2⅜	2½	2⅝	2¾	2⅞	3	
Wt.	lbs. 17.51	lbs. 19.63	lbs. 21.8	lbs. 24.24	lbs. 26.72	lbs. 29.33	lbs. 32.05	lbs. 34.9	

Binary and Decimal Fractions.

$\frac{1}{64}$ =.015625
$\frac{1}{32}$ =.03125
$\frac{3}{64}$ =.046875
$\frac{1}{16}$ =.0625
$\frac{5}{64}$ =.078125
$\frac{3}{32}$ =.09375
$\frac{7}{64}$ =.109375
$\frac{1}{8}$ =.125
$\frac{9}{64}$ =.140625
$\frac{5}{32}$ =.15625
$\frac{11}{64}$ =.171875
$\frac{3}{16}$ =.1875
$\frac{13}{64}$ =.203125
$\frac{7}{32}$ =.21875
$\frac{15}{64}$ =.234375
$\frac{1}{4}$ =.25
$\frac{17}{64}$ =.265625
$\frac{9}{32}$ =.28125
$\frac{19}{64}$ =.296875
$\frac{5}{16}$ =.3125
$\frac{21}{64}$ =.328125
$\frac{11}{32}$ =.34375
$\frac{23}{64}$ =.359375
$\frac{3}{8}$ =.375
$\frac{25}{64}$ =.390625
$\frac{13}{32}$ =.40625
$\frac{27}{64}$ =.421875
$\frac{7}{16}$ =.4375
$\frac{29}{64}$ =.453125
$\frac{15}{32}$ =.46875
$\frac{31}{64}$ =.484375
$\frac{1}{2}$ =.5
$\frac{33}{64}$ =.515625
$\frac{17}{32}$ =.53125
$\frac{35}{64}$ =.546875
$\frac{9}{16}$ =.5625
$\frac{37}{64}$ =.578125
$\frac{19}{32}$ =.59375
$\frac{39}{64}$ =.609375
$\frac{5}{8}$ =.625
$\frac{41}{64}$ =.640625
$\frac{21}{32}$ =.65625
$\frac{43}{64}$ =.671875
$\frac{11}{16}$ =.6875
$\frac{45}{64}$ =.703125
$\frac{23}{32}$ =.71875
$\frac{47}{64}$ =.734375
$\frac{3}{4}$ =.75
$\frac{49}{64}$ =.765625
$\frac{25}{32}$ =.78125
$\frac{51}{64}$ =.796875
$\frac{13}{16}$ =.8125
$\frac{53}{64}$ =.828125
$\frac{27}{32}$ =.84375
$\frac{55}{64}$ =.859375
$\frac{7}{8}$ =.875
$\frac{57}{64}$ =.890625
$\frac{29}{32}$ =.90625
$\frac{59}{64}$ =.921875
$\frac{15}{16}$ =.9375
$\frac{61}{64}$ =.953125
$\frac{31}{32}$ =.96875
$\frac{63}{64}$ =.984375
1 =1.000000

Distances at which to open a 2 ft. Rule to obtain a given Angle.

Angle.	Distance.	Angle.	Distance.	Angle.	Distance.	Angle.	Distance.	Angle.	Distance.
Deg.	Inches	Deg.	Inches	Deg.	Inches	Deg.	Inches	Deg.	Inches
1	.2	19	3.96	37	7.61	55	11.08	73	14.28
2	.42	20	4.17	38	7.81	56	11.27	74	14.44
3	.63	21	4.37	39	8.01	57	11.45	75	14.61
4	.84	22	4.58	40	8.20	58	11.64	76	14.78
5	1.05	23	4.78	41	8.40	59	11.82	77	14.94
6	1.26	24	4.99	42	8.60	60	12.00	78	15.11
7	1.47	25	5.19	43	8.80	61	12.18	79	15.27
8	1.67	26	5.40	44	8.99	62	12.36	80	15.43
9	1.88	27	5.60	45	9.18	63	12.54	81	15.59
10	2.09	28	5.81	46	9.38	64	12.72	82	15.75
11	2.30	29	6.01	47	9.57	65	12.90	83	15.90
12	2.51	30	6.21	48	9.76	66	13.07	84	16.06
13	2.72	31	6.41	49	9.95	67	13.25	85	16.21
14	2.92	32	6.62	50	10.14	68	13.42	86	16.37
15	3.13	33	6.82	51	10.33	69	13.59	87	16.52
16	3.34	34	7.02	52	10.52	70	13.77	88	16.67
17	3.55	35	7.22	53	10.71	71	13.94	89	16.82
18	3.75	36	7.42	54	10.90	72	14.11	90	16.97

French Mètre reduced to Inches.

Mètre.	Inches.		Mètre.	Inches.	Feet.
.001587	= 1/16	**Millimètres** 1 =	.001 =	.03937 =	.00328
.00317	= 1/8	2 =	.002 =	.07874 =	.00656
.00476	= 3/16	3 =	.003 =	.11811 =	.00984
.00635	= 1/4	4 =	.004 =	.15748 =	.01312
.00794	= 5/16	5 =	.005 =	.19685 =	.01641
.00952	= 3/8	6 =	.006 =	.23622 =	.01969
.01111	= 7/16	7 =	.007 =	.2756 =	.02397
.01270	= 1/2	8 =	.008 =	.31497 =	.02625
.01429	= 9/16	9 =	.009 =	.35434 =	.02953
.01587	= 5/8				
.01746	= 11/16	**Centimètres**			
.01905	= 3/4				
.02064	= 13/16	1 =	.01 =	.3937 =	.0328
.02222	= 7/8	2 =	.02 =	.7874 =	.0656
.02381	= 15/16	3 =	.03 =	1.1811 =	.0984
.02540	= 1	4 =	.04 =	1.5748 =	.1312
.05078	= 2	5 =	.05 =	1.9685 =	.1641
.0762	= 3	6 =	.06 =	2.3622 =	.1969
.1016	= 4	7 =	.07 =	2.756 =	.2397
.1270	= 5	8 =	.08 =	3.1497 =	.2625
.1524	= 6	9 =	.09 =	3.5434 =	.2953
.1778	= 7				
.2032	= 8	**Decimètres**			
.2286	= 9				
.2540	= 10	1 =	.1 =	3.9371 =	.3281
.2794	= 11	2 =	.2 =	7.8742 =	.6562
.3048	= 12	3 =	.3 =	11.8112 =	.9843
		4 =	.4 =	15.7483 =	1.3124
		5 =	.5 =	19.6854 =	1.6404
		6 =	.6 =	23.6225 =	1.9685
		7 =	.7 =	27.5596 =	2.3966
		8 =	.8 =	31.4966 =	2.6247
		9 =	.9 =	35.4337 =	2.9528

THE MÈTRE = 3.2808992 FEET (ABOUT 39⅜ INCHES).

PLATES OF GEAR TEETH

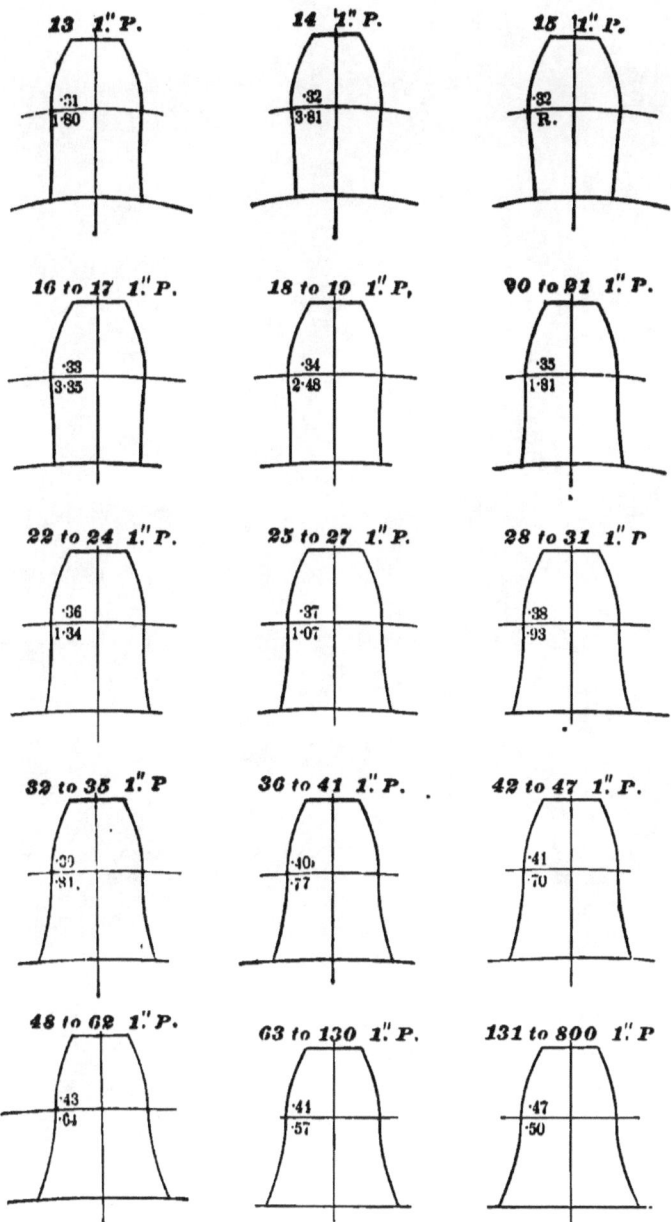

FULL SIZE GEAR TEETH.
From Prof. S. W. Robinson's Templet Odontograph.

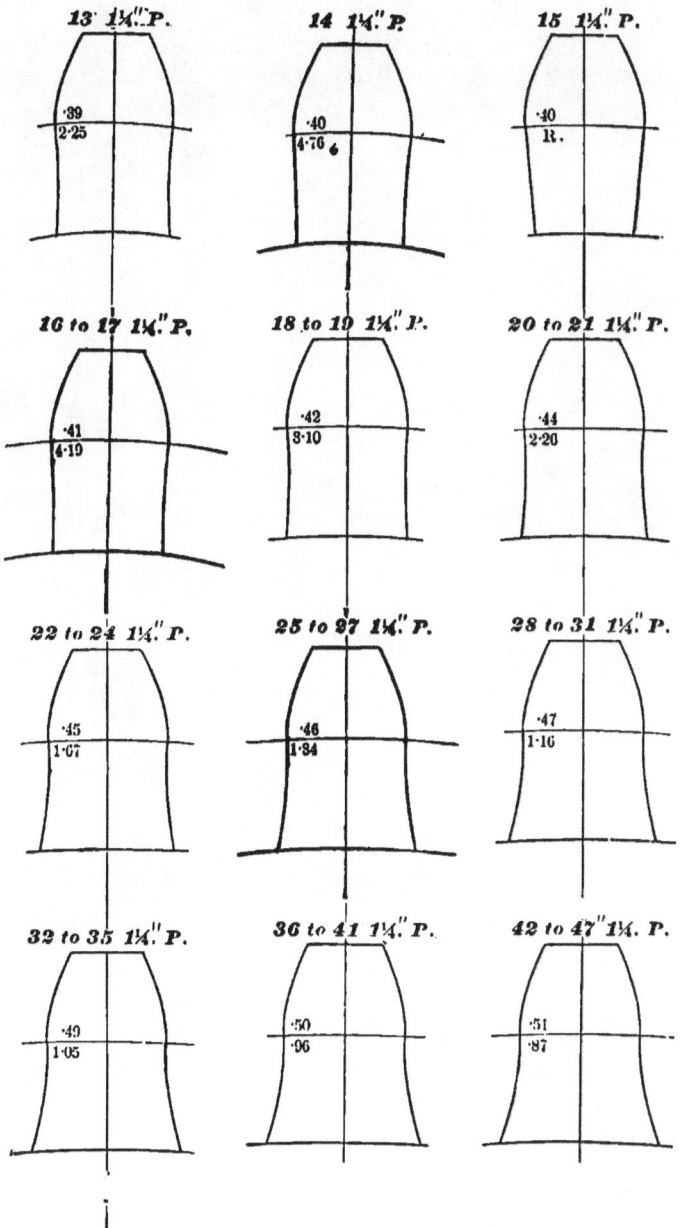

FULL SIZE GEAR TEETH.
From Prof. S. W. Robinson's Templet Odontograph.

CALIFORNIA

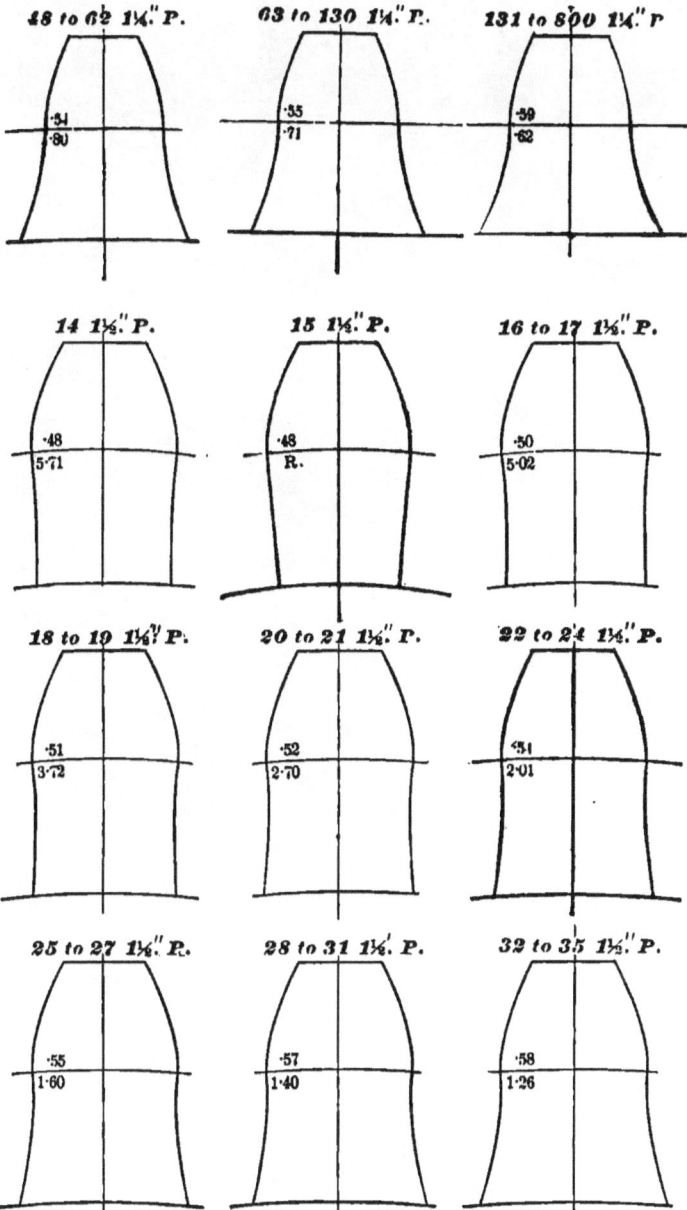

FULL SIZE GEAR TEETH.
From Prof. S. W. Robinson's Templet Odontograph.

CALIFORNIA

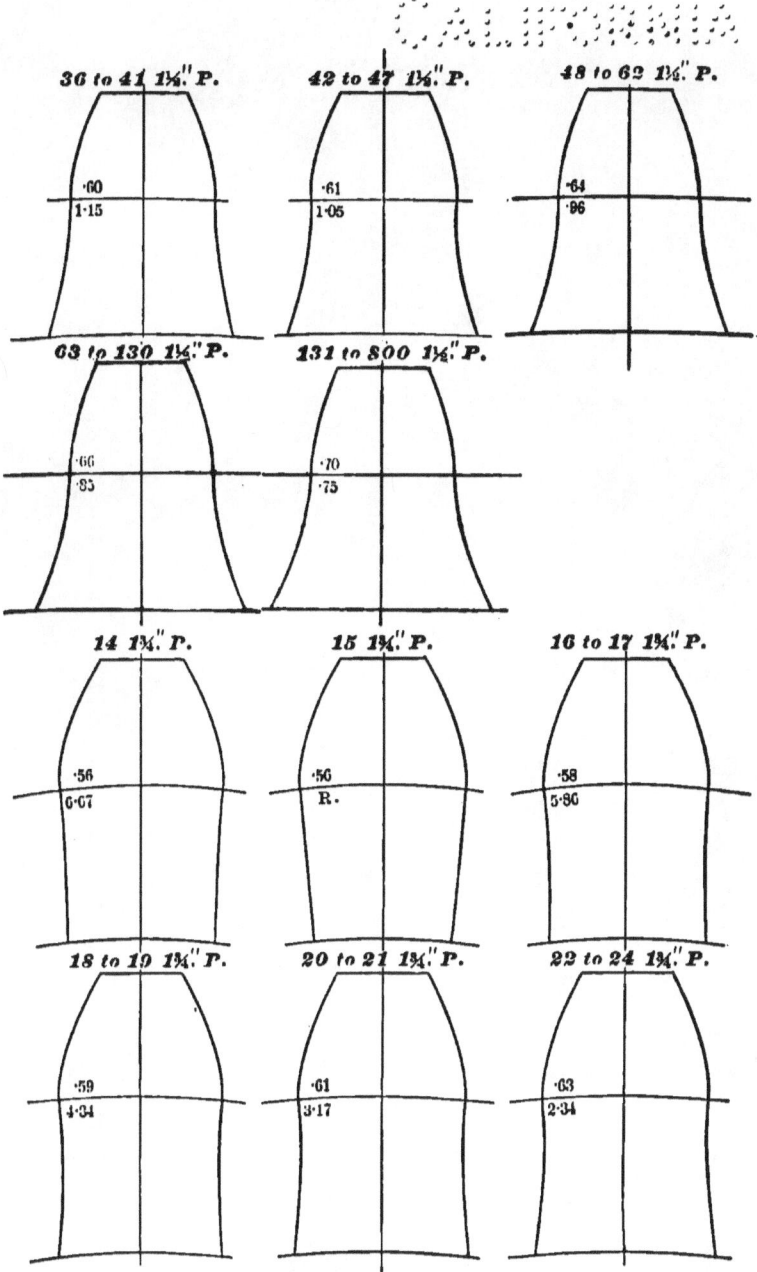

FULL SIZE GEAR TEETH.
From Prof. S. W. Robinson's Templet Odontograph.

UNIV. OF
CALIFORNIA

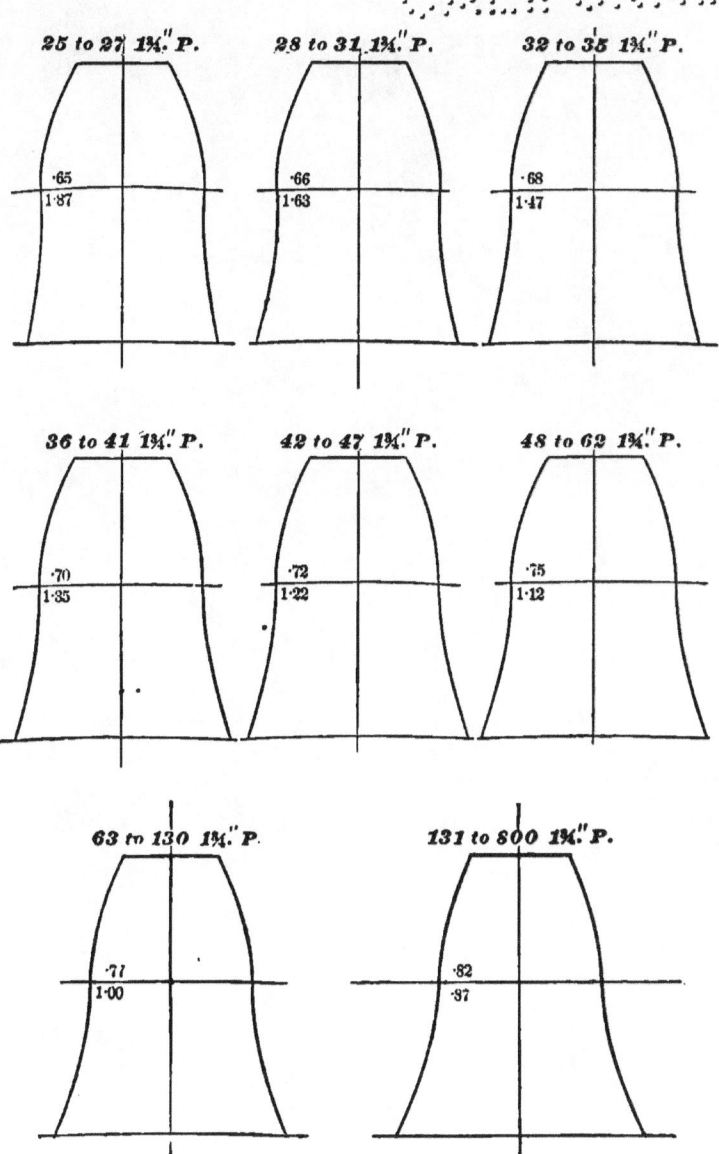

FULL SIZE GEAR TEETH.
From Prof. S. W. Robinson's Templet Odontograph.

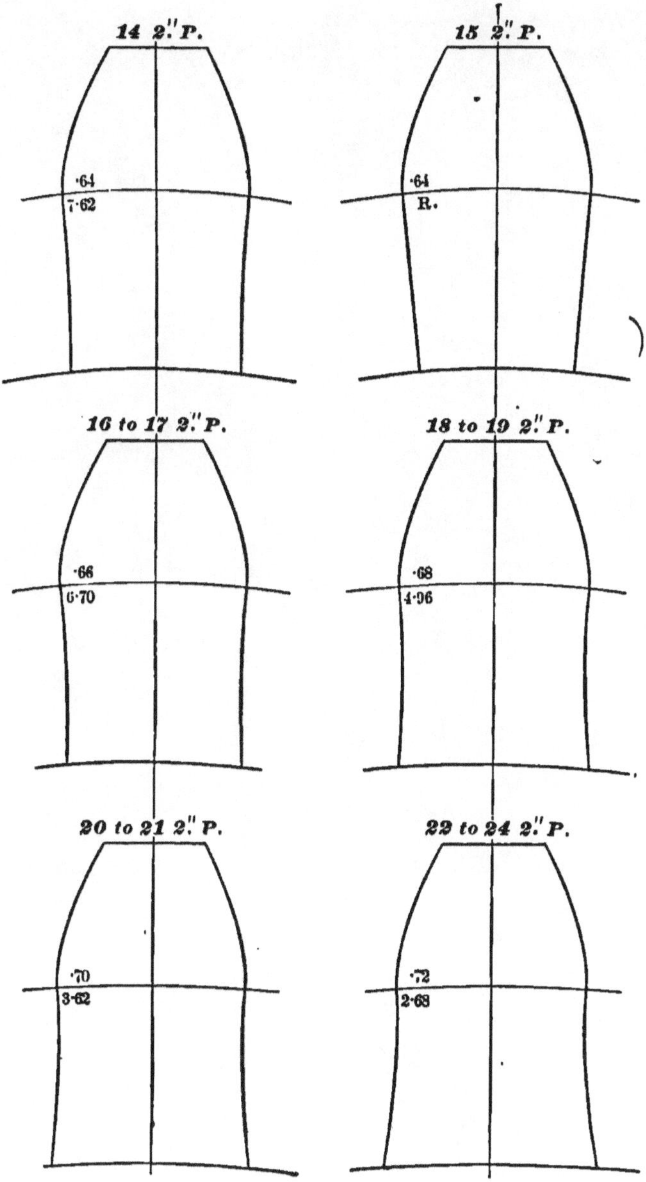

FULL SIZE GEAR TEETH.
From Prof. S. W. Robinson's Templet Odontograph.

Univ. of
California

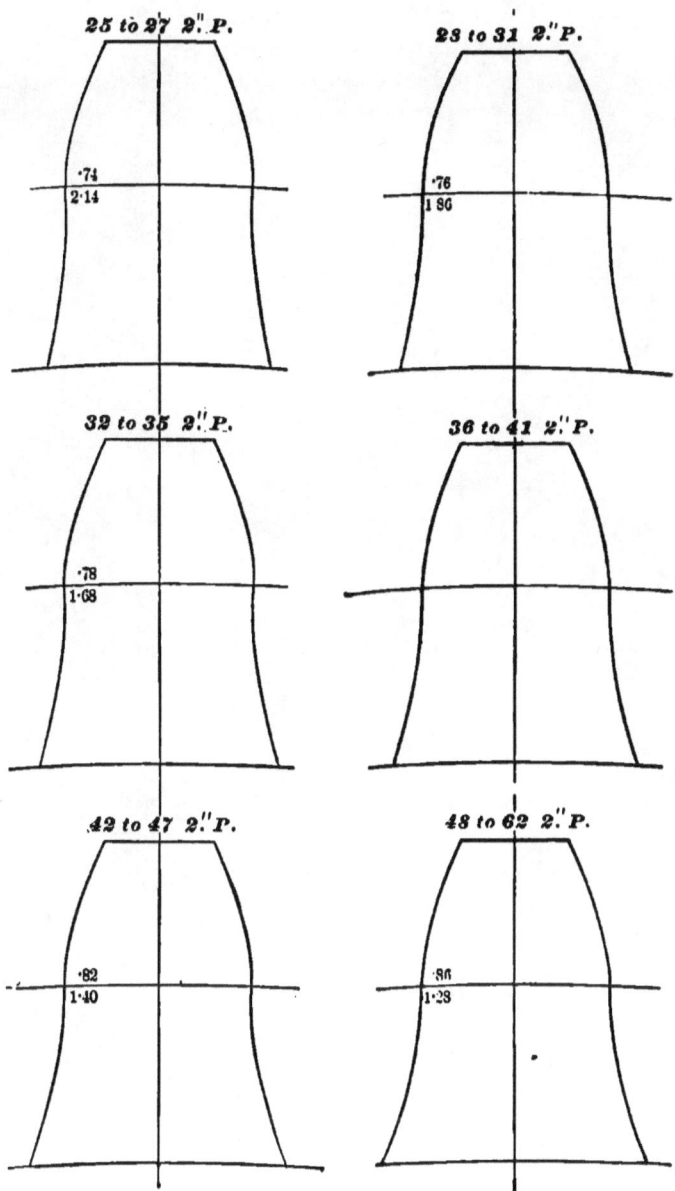

FULL SIZE GEAR TEETH.
From Prof. S. W. Robinson's Templet Odontograph.

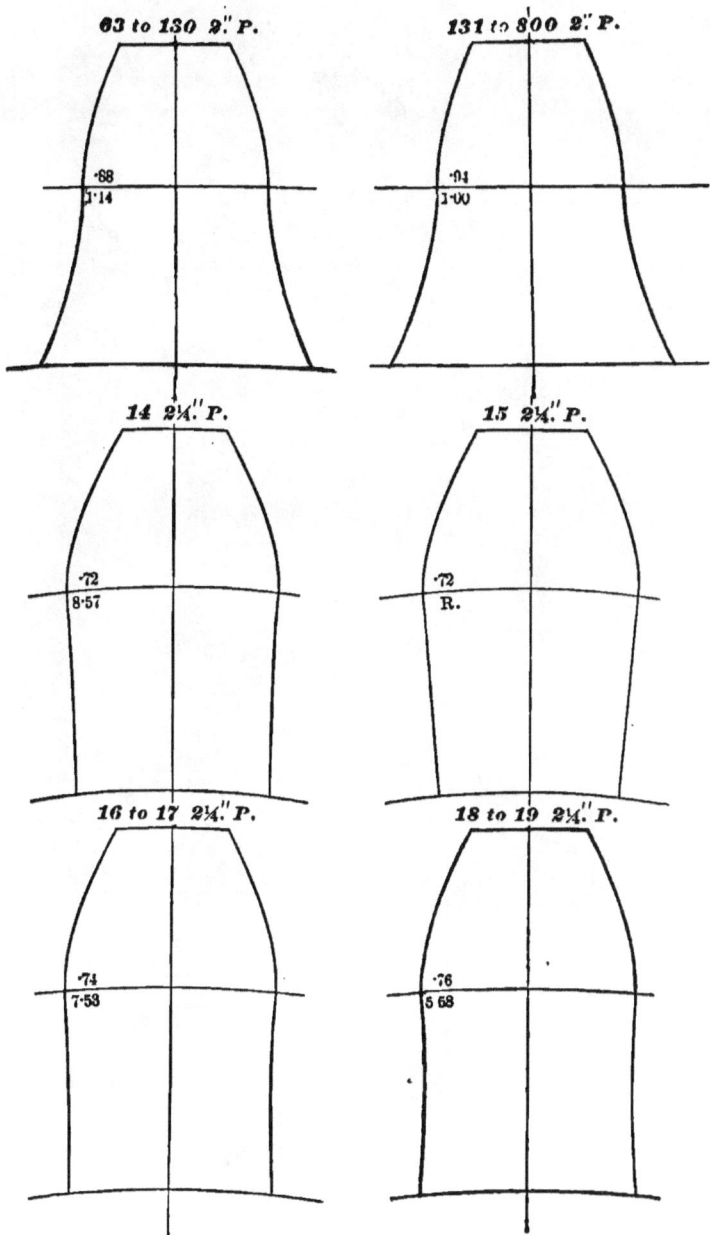

FULL SIZE GEAR TEETH.
From Prof. S. W. Robinson's Templet Odontograph.

CALIFORNIA

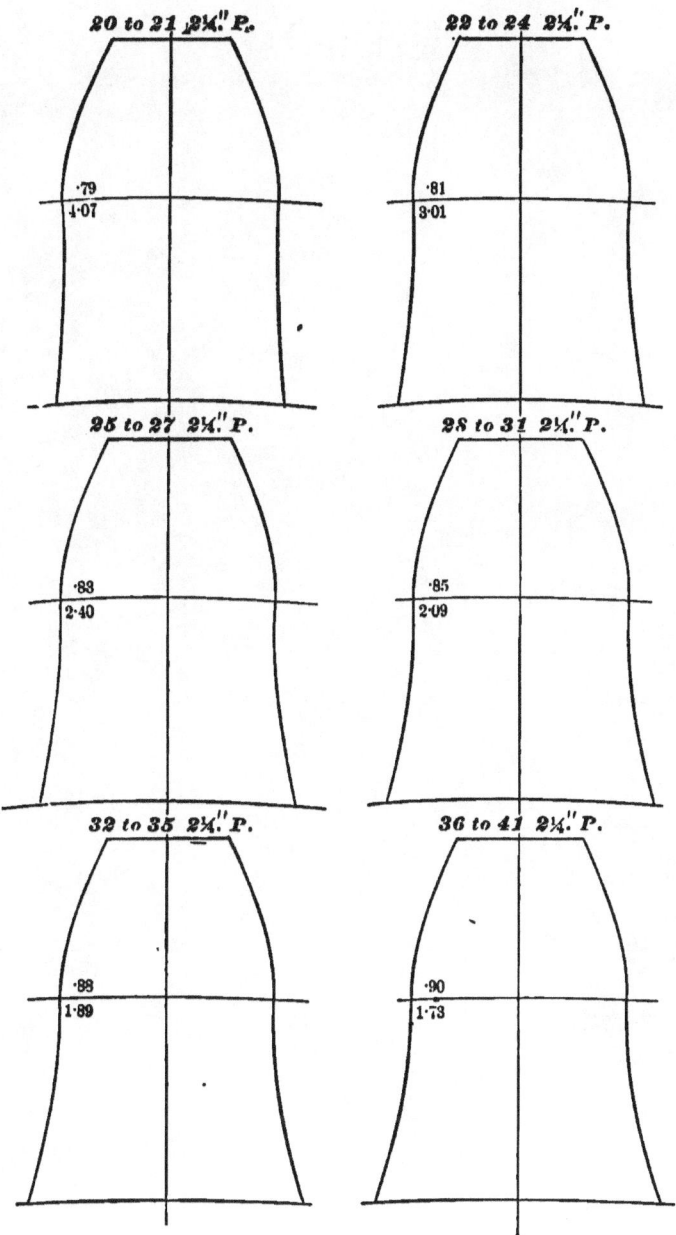

FULL SIZE GEAR TEETH.
From Prof. S. W. Robinson's Templet Odontograph.

CALLIOPE

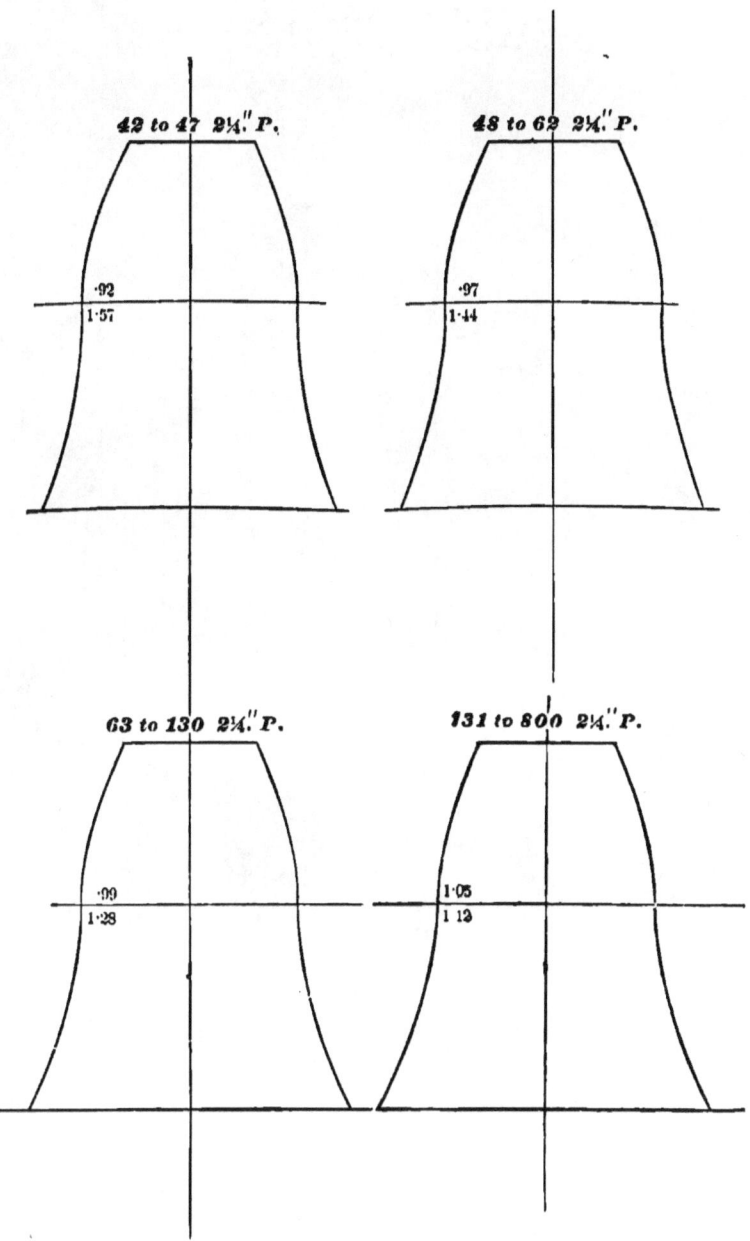

FULL SIZE GEAR TEETH.
From Prof. S. W. Robinson's Templet Odontograph.

CALIFORNIA

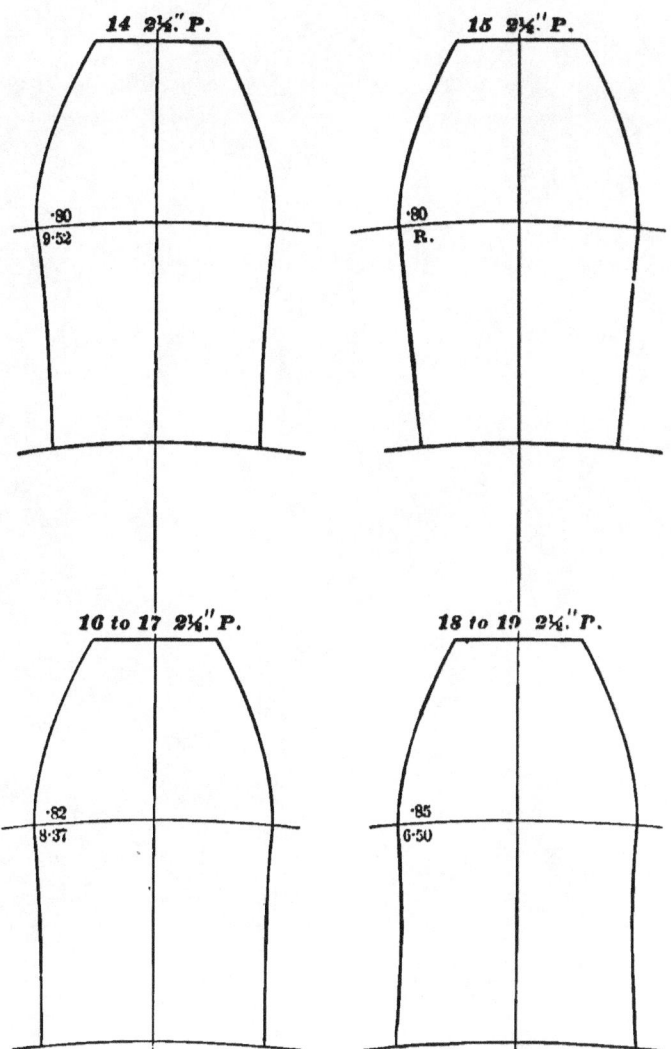

FULL SIZE GEAR TEETH.
From Prof. S. W. Robinson's Templet Odontograph.

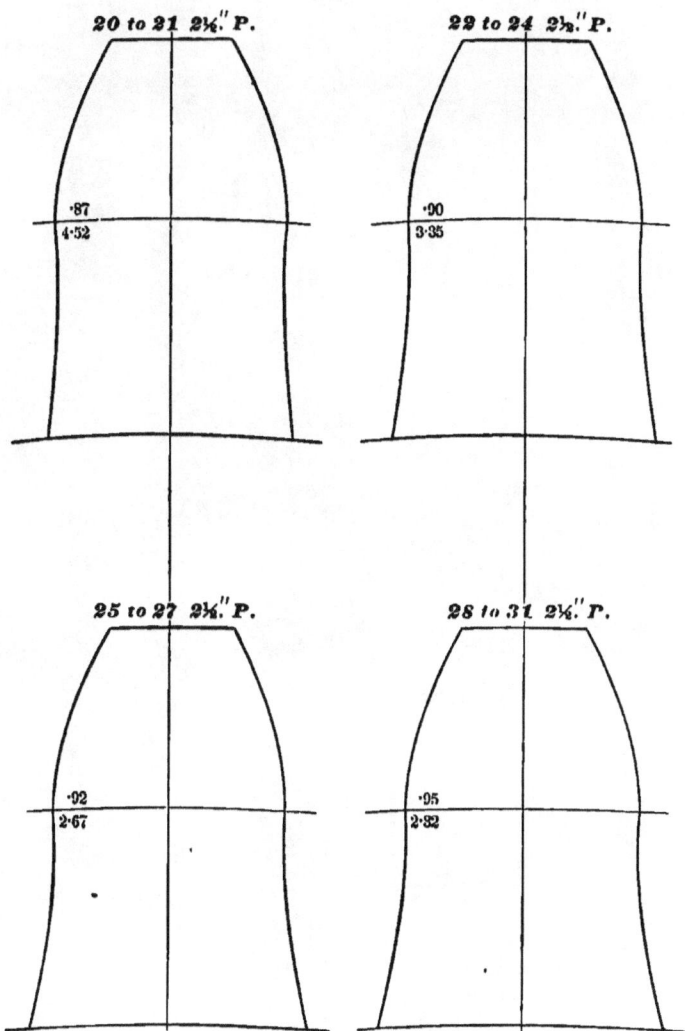

FULL SIZE GEAR TEETH.
From Prof. S. W. Robinson's Templet Odontograph.

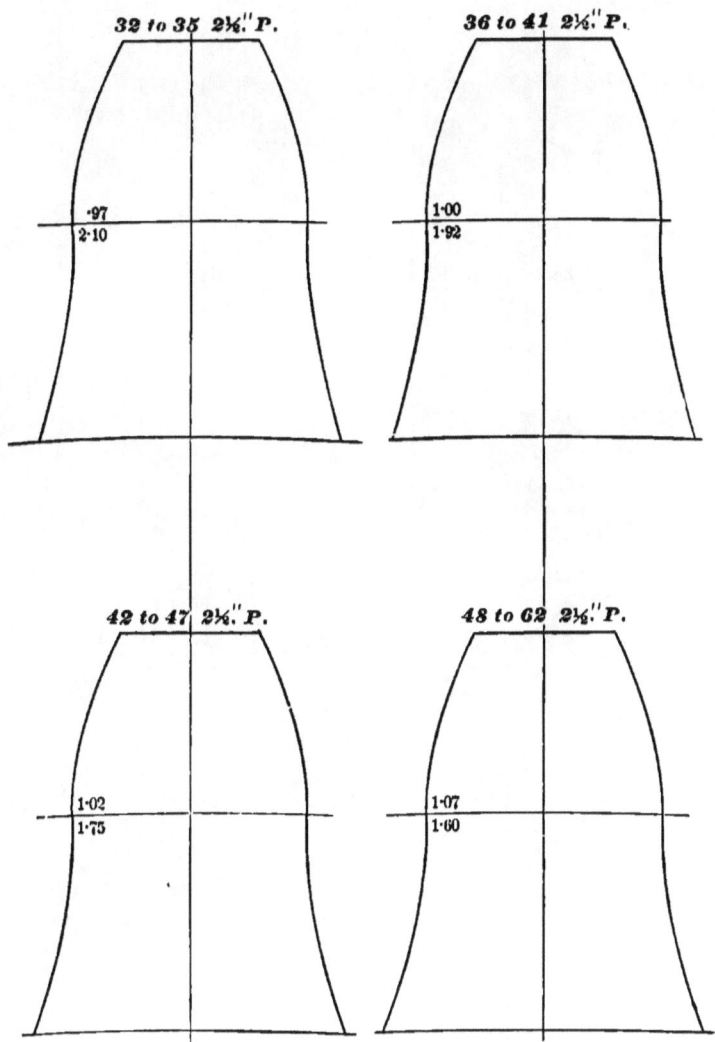

FULL SIZE GEAR TEETH.
From Prof. S. W. Robinson's Templet Odontograph.

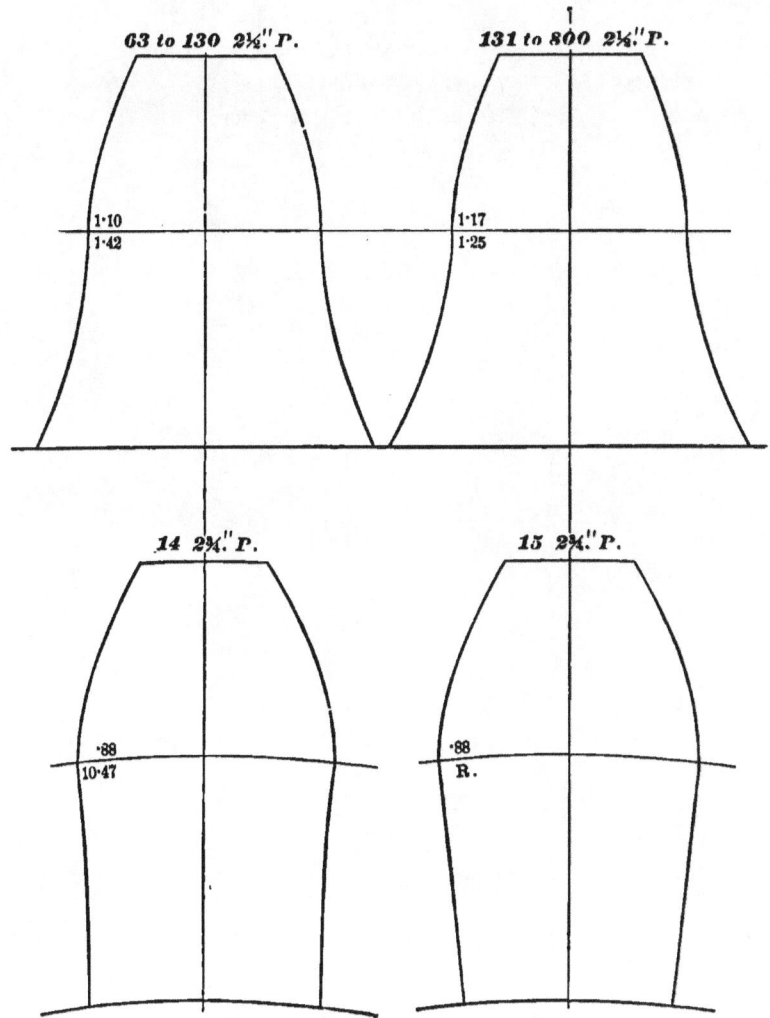

FULL SIZE GEAR TEETH.
From Prof. S. W. Robinson's Templet Odontograph.

UNIV. OF
CALIFORNIA

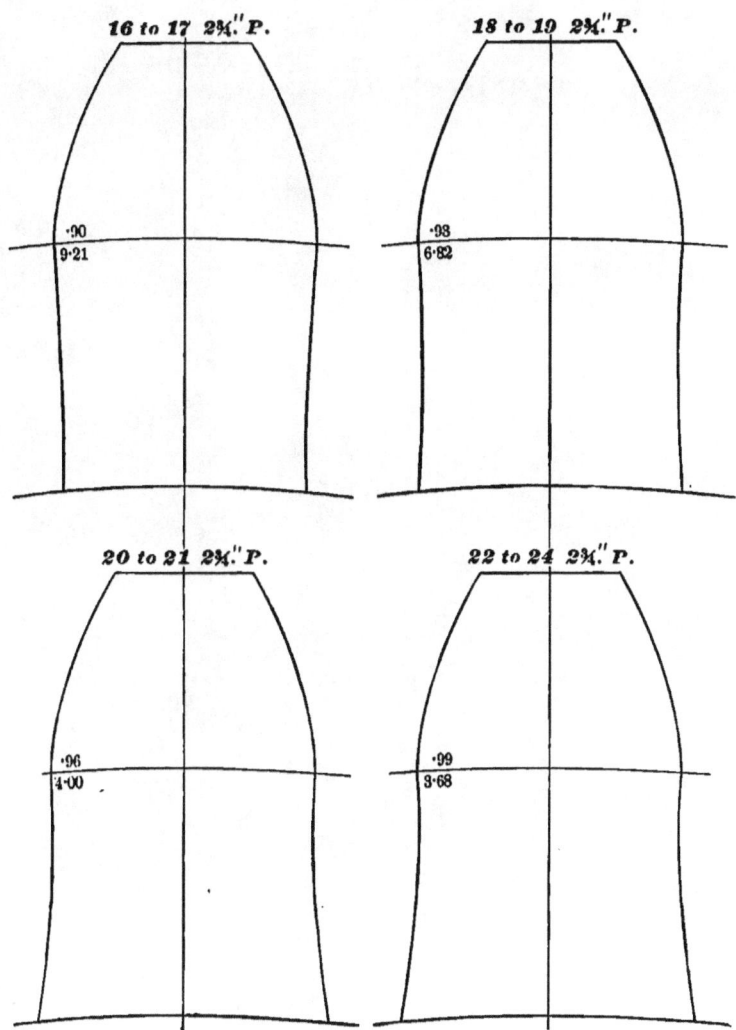

FULL SIZE GEAR TEETH.
From Prof. S. W. Robinson's Templet Odontograph.

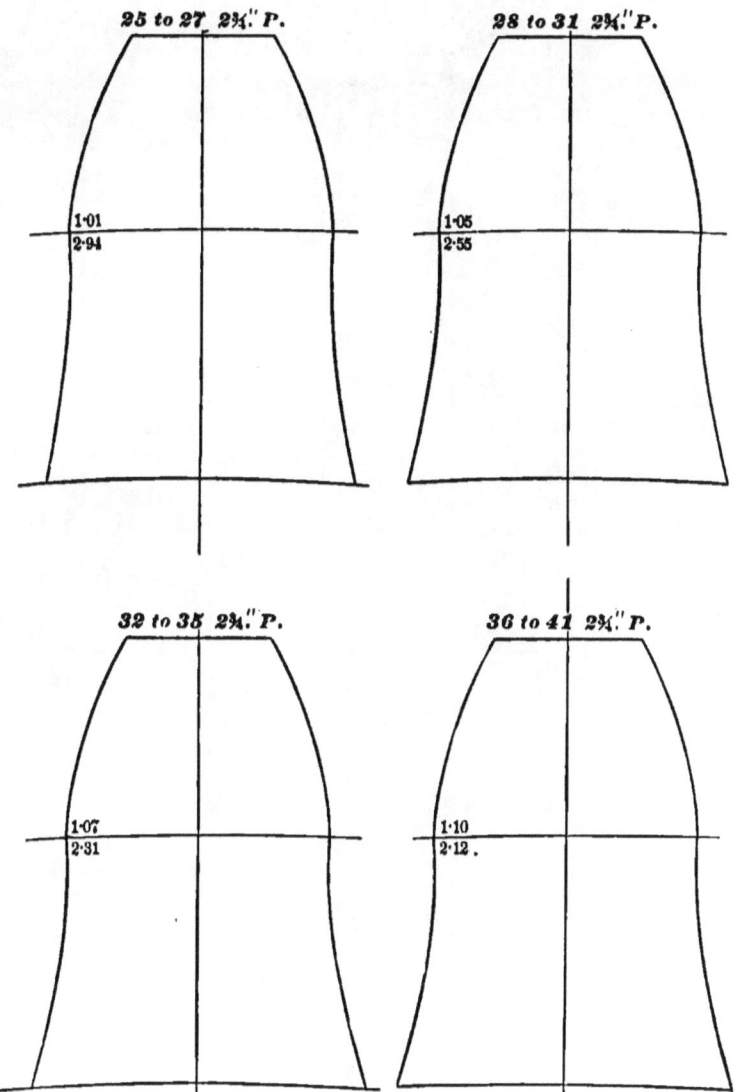

FULL SIZE GEAR TEETH.
From Prof. S. W. Robinson's Templet Odontograph.

UNIV. OF
CALIFORNIA

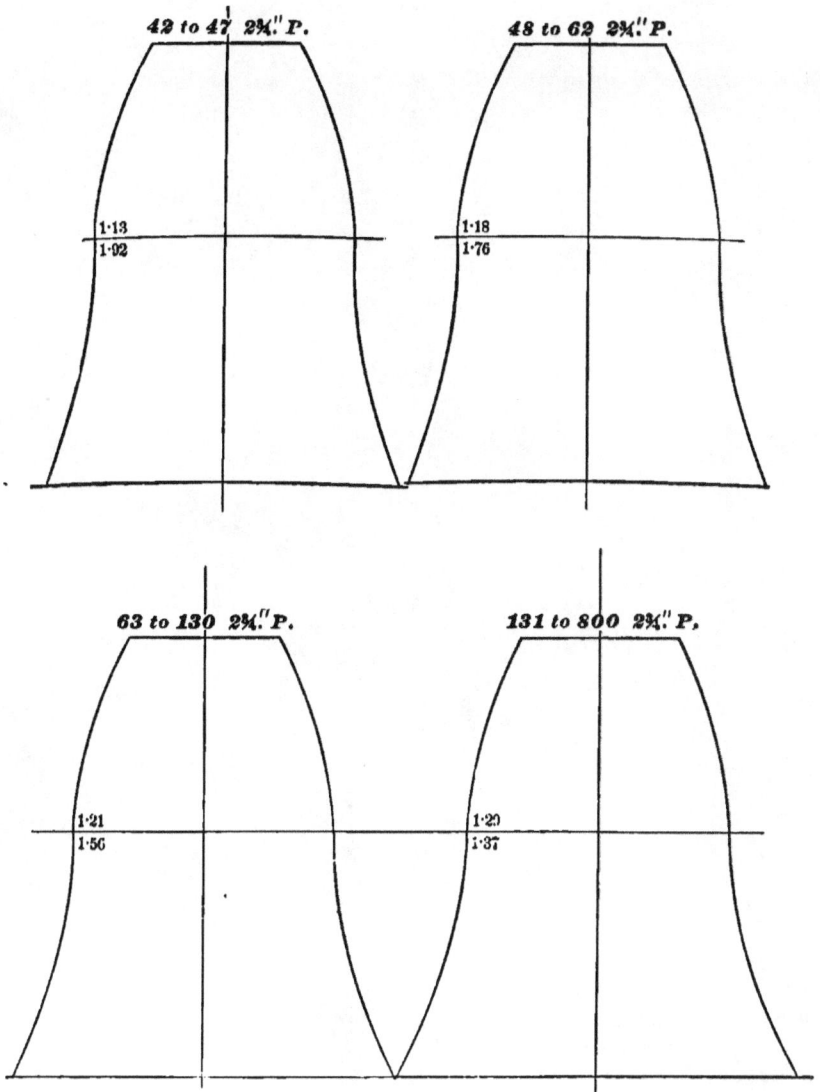

FULL SIZE GEAR TEETH.
From Prof. S. W. Robinson's Templet Odontograph.

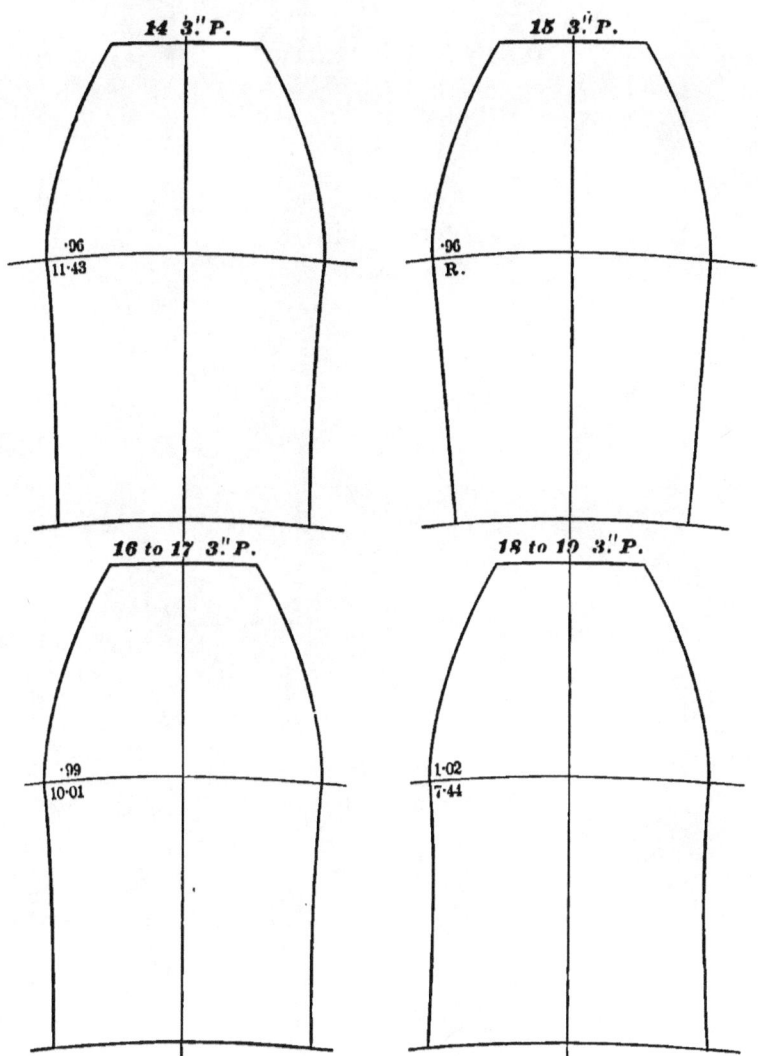

FULL SIZE GEAR TEETH.
From Prof. S. W. Robinson's Templet Odontograph.

CALIFORNIA

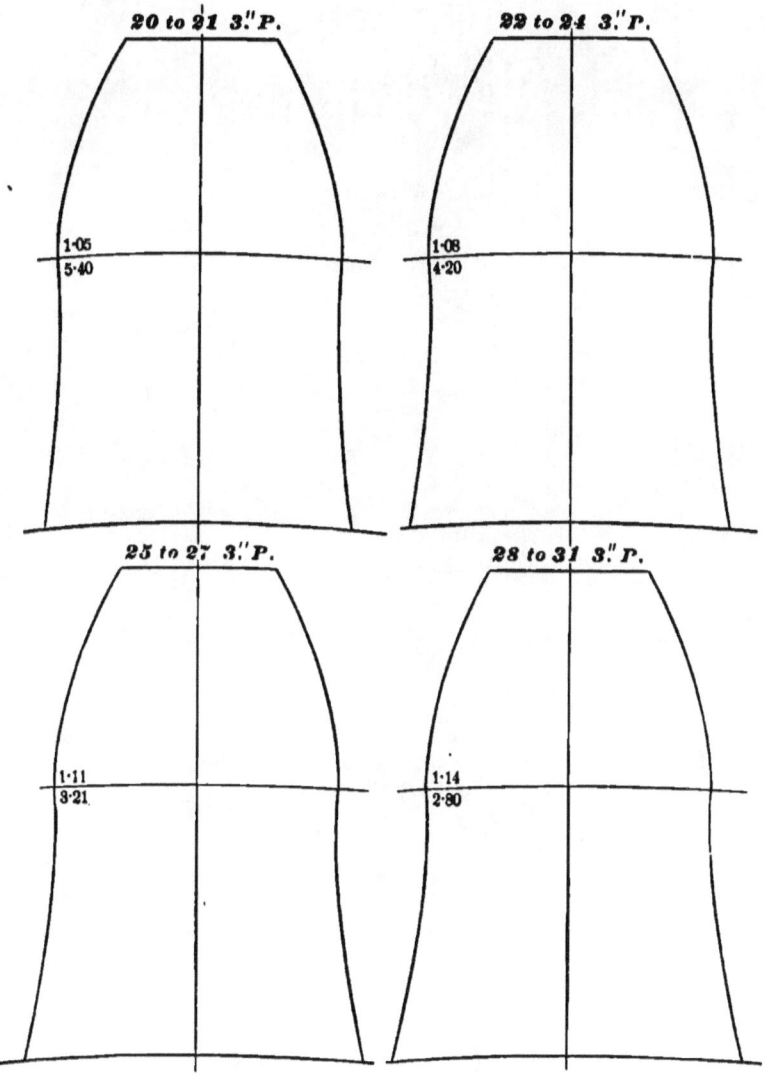

FULL SIZE GEAR TEETH.
From Prof. S. W. Robinson's Templet Odontograph.

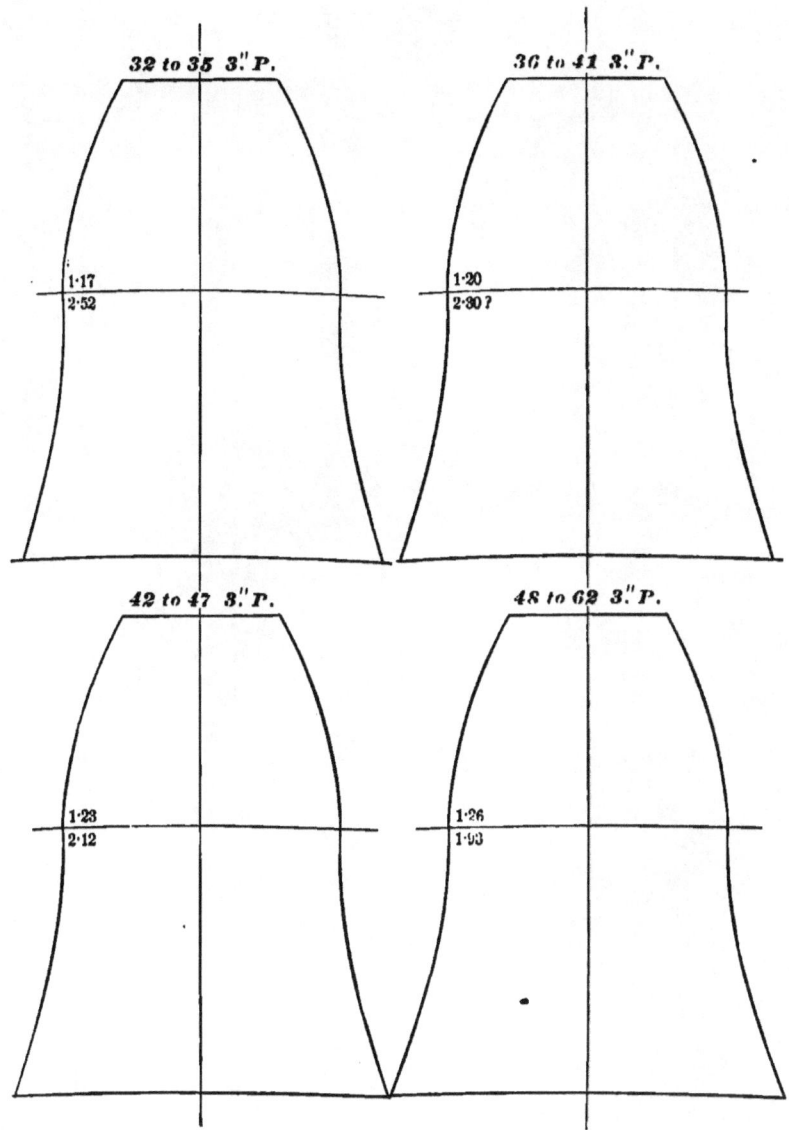

FULL SIZE GEAR TEETH.
From Prof. S. W. Robinson's Templet Odontograph.

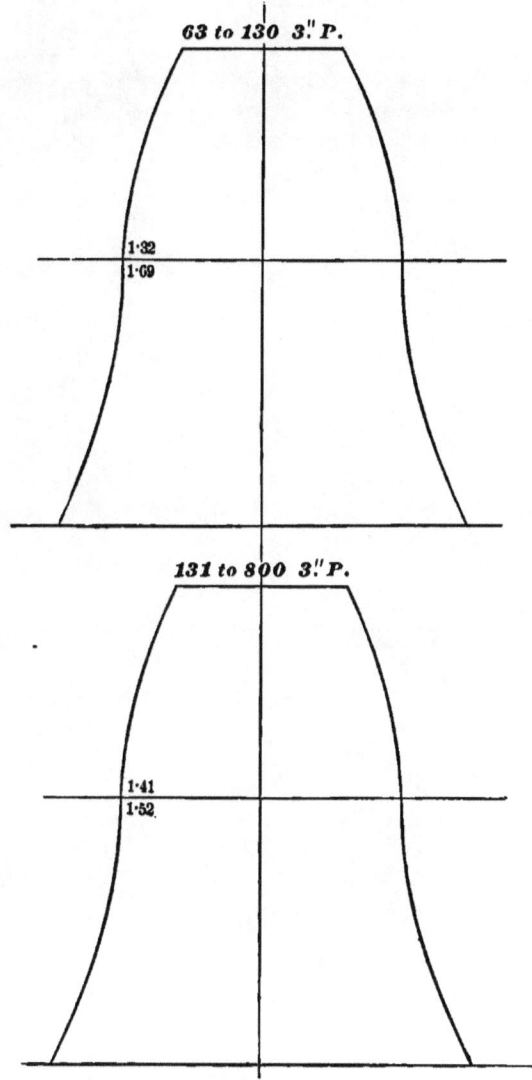

FULL SIZE GEAR TEETH.
From Prof. S. W. Robinson's Templet Odontograph.

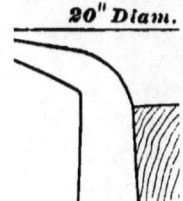

www.ingramcontent.com/pod-product-compliance
Lightning Source LLC
Chambersburg PA
CBHW020248170426

43202CB00008B/274